VEGETATION DYNAMICS

VEGETATION DYNAMICS

Proceedings of the Second Symposium of the Working Group
on Succession Research on Permanent Plots, held at
the Delta Institute for Hydrobiological Research,
Yerseke, October 1–3, 1975

edited by

Wim G. Beeftink

Dr. W. Junk bv Publishers   The Hague – Boston – London 1980

ISBN 90 6193 606 3

CONTENTS

VEGETATION DYNAMICS IN RETROSPECT AND PROSPECT INTRODUCTION TO THE PROCEEDINGS OF THE SECOND SYMPOSIUM OF THE WORKING GROUP ON SUCCESSION RESEARCH ON PERMANENT PLOTS

Wim G. BEEFTINK*

Delta Institute for Hydrobiological Research, Yerseke, The Netherlands**

Keywords:
Geobotany, Paradigm, Permanent plots, Succession, Symposium, Vegetation dynamics

The Working Group on Succession Research on Permanent Plots was erected at the 17th Symposium of the International Society for Vegetation Science, held at Rinteln (F.R.G.), Spring 1973. The immediate cause for the start of this group was the presentation of some lectures on succession research, especially that of H. Böttcher (Hannover) dealing with the actual position of this kind of research in European countries (Schmidt 1974, Böttcher 1975). The organization of the Working Group, including the animation of scientific meetings and excursions, has been laid (and is still) in the hands of Dr. Wolfgang Schmidt, Systematic-Geobotanical Institute, The University, Untere Karspüle 2, D3400-Göttingen (F.R.G.).

An inventory of succession research in West-European countries carried out by the above-mentioned colleagues, proved that most work is done in the Netherlands. Today, ca. 3000 permanent plots are or have been studied by workers from various Dutch ecological institutions, some of them already from before World War II (Research Institute for Nature Management, Leersum, Annual Reports 1974–77; Beeftink 1975). Equally long, and even longer, are the experiences obtained in Great Britain (Watt 1960, 1971, Thurston et al. 1976, Williams 1978). Other countries from which research studies on permanent plots are known, are the Federal Republic of Germany, Poland, Sweden and Switzerland.

The inventory also showed that ecologists are especially interested in the dynamics of grasslands and similar vegetation types such as salt marshes and dune slacks. Of second interest are those of abandoned agricultural fields and of forests and brushwood, including clearings. Minor attention has been payed to the dynamics of dry dune, moor and heath vegetation and aquatic and freshwater littoral vegetation. Most of these studies appear to be connected with problems concerning ecological side-effects of land and water management and of civil engineering works, and rarely with problems of pure scientific and theoretical interest.

The first meeting of the Working Group was held at Rinteln, 10 April 1974, during the 18th Symposium of the International Society. Two themes were discussed: (a) Vegetation development in abandoned agricultural fields and the influence of management measures on it, and (b) the most appropriate methods of vegetation analyses to be used (Schmidt 1974). The papers were united with those of the Proceedings of the 17th Symposium under the title: Succession Research (Sukzessionsforschung) (Schmidt 1975). The members of the Group considered it desirable to present results and discuss problems in succession research from both practical and theoretical points of view.

For a second meeting in 1975 the Working Group was invited by the Delta Institute for Hydrobiological Research, Yerseke, the Netherlands. Participants were requested to deliver case studies of research on permanent plots. The results are presented in this volume.

Different approaches and object studies have been presented at this meeting. Londo emphasized once more the technique and method of vegetation analysis on permanent plots. Thalen used aerial photographic techniques in permanent plot studies. Faliński gathered trans-

* The author is much indebted to Dr. A.H.L. Huiskes and Dr. K.F. Vaas (Yerseke) and to Dr. E. van der Maarel (Nijmegen) for criticizing and reviewing the text.
** Communication Nr. 185.

formations in a Polish primeval forest from a 10 year pattern of decay of uprooted trees. Romane (see Poissonet et al.) reported the results of experiments in a *Quercus coccifera* garrigue in order to elucidate the impact of man's action. Isépy showed the value of transitional zones between transdanubian grassland and forest for mountain and subalpine relict plants. Bråkenhielm reported on his results on vegetion dynamics in afforested farmland in Sweden and Schmidt studied nitrogen mineralization in an old agricultural field. Van der Maarel presented preliminary results of experimental succession research in a Dutch coastal dune grassland. Van der Laan described spatial and temporal variations in dune-slack vegetation in relation to the ground water régime.

Van Noordwijk-Puijk et al. reported on their numerical pattern detection studies in the vegetation of sand flats isolated from tidal action, and Joenje described the migration and colonization processes of plants in such mud flats. Schmeisky communicated on his succession research in a Baltic salt marsh and Bakker on his management experiments in a Wadden Sea salt marsh.

Muhle, Nienhuis, Polderman and Simons introduced a separate group of object studies: the results and technical problems of permanent plot studies in cryptogamic communities. Bilio communicated on his problems with experimental panels for evaluating fouling in marine environments. Finally, Krause gave a survey of the succession research carried out by the German (F.R.G.) State Institute for Nature Conservation and Landscape Ecology.

In the meantime a third meeting took place, from 30 August to 2 September 1977, in the Białowieza Geobotanical Station of the Warsaw University, Poland (Werger & Schmidt 1978). The text of the papers presented will be published in Phytocoenosis, the phytosociological bulletin of the University of Warsaw, under the editorship of Dr. J. B. Faliński.

It is intriguing to consider the status of the study of vegetation dynamics from the view-point of historical and future development of vegetation science and ecosystem research. Firstly, it can be stated that, although vegetation dynamics was considered as an essential part in the Braun-Blanquet approach right from the beginning (Braun-Blanquet 1928, Lüdi 1930), the temporal dimension in the study of vegetation received little attention in practice, in fact less than in the Anglo-American (Clements 1916, Kershaw 1973) and Russian approaches (see e.g. their contributions in Knapp's (1974) book on vegetation dynamics). This discrepancy can be explained from the fact that

vegetation scientists following the Braun-Blanquet approach, had invested much energy in questions of the syntaxonomy and synecology of vegetation types. This preoccupation may be understood partly from scientific leadership stressing the description and floristic characterization of vegetation types on national and supranational levels. For another part, however, it may be more a matter of generally accepted and emphasized paradigms. Simberloff (1976), citing Kuhn (1970), defines a paradigm as a theory so widely accepted as an accurate description of nature that failure of an experiment to yield the result deduced from the theory leads not to rejection of the theory but rather to attemps to find fault with the deductive logic or experimental procedure, or simply to willful suspension of believe in the experimental result. Especially the holistic postulates concerning the organismic character of communities (the whole is more than the sum of the parts) and of the floristic and structural uniformity under certain environmental conditions (cf. Westhoff & Van der Maarel 1978), must be mentioned in this connection.

It is not my intention to discuss criticisms on the Braun-Blanquet approach (see for that e.g. Whittaker 1962, Westhoff & Van der Maarel 1978), nor to evaluate the theoretical implications of the underlying hypotheses and statements in vegetation science and ecology (who will take up this task?), but only to signalize some trends in order to find reasons for the present state of vegetation dynamics and for its future development.

There are signs of an increased interest in vegetation dynamics amongst continental ecologists in the last two decades. Three groups of reasons can be mentioned (see also Knapp 1974):

(a) In some West-European countries the classification of communities is fading out as a research topic or has almost reached saturation. This has different grounds: One is that much of the natural and semi-natural vegetation is degenerating and even deteriorating in those countries owing to human pressure (Van der Maarel 1975, Westhoff 1976), that further work becomes problematical. Another ground is that the scientific need for further refinements of the existing classification system is decreasing. Only a more objective insight into vegetational relationships on a supranational or even a continental scale, by means of numerical processing of extensive data sets, has still the interest of some groups of scientists (Van der Maarel 1974, Van der Maarel et al. 1976). For that reason continental vegetation scientists are beginning to look for alternative lines of research within their discipline which will open new perspectives in understanding structure and function in ecological relationships. One of these lines is the study of vegetation structure and vegetation morphology. Such a study has to be related (in a both descriptive and experimental way) to the morphology and strategy of plants as well as to environmental stress in order to give an ecological and functional interpretation of the architecture of plant communities and of phenomena of structural convergence and divergence, comparing the vegetation in different parts of the world (e.g. Pümpel 1977, Werger & Ellenbroek 1978, Shmida & Whittaker in press).

Another line of research tends to reduce the problems to those of lower levels of organization: the research of the dynamics and genetics of plant populations within communities (e.g. Beeftink et al. 1978, Beeftink 1979). A third one is concerned with interactions between trophic levels, and with flow and storage (turnover) of organic matter and nutrients, as well as of materials added anthropogenically to the ecosystem (e.g. Teal 1962, Odum 1971, Tyler 1971, Ketner 1972, Banus et al. 1975, Valiela et al. 1975). In all these lines of research the temporal aspects of the object of study are an essential part of examination.

(b) The degradation of vegetation as a consequence of man's impact on Nature becomes more and more problematic. Nature conservation and nature management issues arising from these changes, urgently call for more information on the relations between vegetational and environmental dynamics. On the basis of experiences gained from nature management in the Netherlands and from plant population studies on permanent plots, Van Leeuwen (1966, 1973, see also Van der Maarel 1976) developed his Relation Theory. In this theory Van Leeuwen formulated basic relations in space and time between internal and external influences operating in a system. This theory has proved to be very useful in practical nature management and in gaining more insight in the ecological backgrounds of man's activities (Westhoff 1970). As a basis for scientific research, however, the theory needs to be further examined. Some hypotheses involved in this theory, especially those concerning the quantifiability of the amount of environmental "uncertainty" or "precariousness" as far as relevant for the behaviour of the organisms or communities involved, have still to be tested on their falsifiability (cf. Van der Steen 1973). We must try not to stick to theories which we accept as an accurate description of nature, but which are in fact clusters of nonoperational assumptions based on a purely logical system (Peters 1976). Van der Maarel & Dauvellier (1978) diagnosed this tautological strain for the Relation Theory of Van Leeuwen. Although its basic conceptions are meaningful for ecological appro-

aches, the system of deductions built up by Van Leeuwen has hitherto been tested empirically in only a few cases (e.g. Strijbosch 1976). Therefore, Van der Maarel & Dauvellier (1978) advocated to specify the Relation Theory into a number of more concrete, falsifiable hypotheses with respect to spatial and temporal relations in ecosystems. This has already been understood by some Dutch ecologists who started a methodologic analysis of the theory (Reddingius & Sloep, University of Groningen) and a pattern analysis in vegetation and environment for testing the theory on gradient situations (Westhoff, Van der Maarel & Roozen, University of Nijmegen).

Such theoretical and ecological consequences have to be worked out urgently, because only such rigid arguments seem to be convincing ultimately as a reliable scientific basis for nature conservation measures, alternative agricultural practices, etc., and, especially, as a guide for environmental politics. However, one comment has to be made on this. In my – and other ecologists' – opinion the degradation and deterioration of nature are going on in such a rapid way (tropical rain forests!) that time will be too short for even a concise scientific foundation of ecological theories. In such a case ecological tautologies may then be useful aids in the deductive work in behalf of nature conservation and environmental politics when accepted as alternatives of true theories. Van der Maarel & Dauvellier (1978) stressed this use especially for the Relation Theory, referring to Peters (1976), however without neglecting the need of empirical testing of hypotheses. Peters (1976) indeed emphasized the restrains of ecological tautologies in use to ecologists dealing with empirical problems. But he is convinced that their utility would be better appreciated if we recognize their logical nature more explicitly.

(c) The modern technical possibilities of instrumental equipment, numerical processing of large data sets and use of dynamic models provide the ecologist with tools for enlarging his choice and accuracy of measuring biotic and abiotic variables considerably, and for applying more sophisticated methods, both in the field and under laboratory conditions. These trends are most promising for finding new ecological aspects and for testing old and new hypotheses for causal and relational explanations of vegetation dynamics and ecosystem behaviour.

After enumerating the reasons for the increase of interest in research by means of permanent plots, transects, grids, etc., it seems useful to consider some ecological problems for which this method will be valuable or even an essential approach. I will touch on five groups of problems, all with both a more theoretical and a more practical bearing:

On the ecosystem level:

(1) The bioenergetic basis for ecological succession in developing ecosystems: how can this be expressed and quantified by structural and functional characteristics as listed by Odum (1969, 1971), criticized by e.g. Orians (1975) and tested by e.g. Bakelaar & Odum (1978).

(2) The interference of the frequency and nature of environmental disturbances, including perturbations brought about by agricultural techniques, with the bioenergetics of different types of ecosystems.

On the community level:

(3) The recognition of spatial and temporal patterns in biological and environmental variables: to which extent a congruency within and between these two groups of variables can be detected in order to find causal and relational explanations for ecological phenomena (Hogeweg 1976, Van Noordwijk-Puijk et al. 1979).

(4) The spatial and temporal relations required for the development of plant and animal communities: how can these relations be expressed by the concepts of diversity and stability (Margalef 1968, Odum 1969, Orians 1975, Whittaker 1975), and which spatial and temporal conditions do those communities need for their survival (Van Leeuwen 1966, Van der Maarel 1971, 1976).

On the population level:

(5) The way biological and ecological mechanisms govern population succession in natural and semi-natural communities, and the role these mechanisms play in community stability and organization (MacArthur & Wilson 1967, Horn 1976, Connell & Slatyer 1977, Grime 1978).

(6) Population regulation within species of natural and semi-natural communities: how to preserve characteristic and rare species, considering their potentials for survival against biotic and abiotic influences (e.g. Harper 1977).

It is not easy to predict further developments in phytosociology. However, one thing seems sure: The present situation of the Braun-Blanquet approach finds itself at a crossing of roads (cf. Pignatti 1975): the classic approach preferably extended with the possibilities of numerical technics, will still need its students to fill up the gaps in our syntaxonomical and synecological knowledge, and to bring hitherto unexplored countries and continents in a large-scale floristic typology of plant communities. On the other hand, underexplored approaches such as vegetation dynamics and research of vegetation structures, and even rather new ways, such as those bridging phytosociology towards other levels of organization (population, ecosystem), will promote more pragmatic lines of research,

borrowed from other geobotanical branches. These latter developments are promising for the ultimate unification of ecological approaches and concepts (cf. Van Dobben & Lowe – McConnell 1975) in the different lines of geobotanical research towards one ecological theory.

References

Bakelaar, R.G. & E.G. Odum. 1978. Community and population level responses to fertilization in an old-field ecosystem. Ecology 59: 660–665.

Banus, M.D., I. Valiela & J.M. Teal. 1975. Lead, zinc and cadmium budgets in experimentally enriched salt marsh ecosystems. Estuar. Coastal Marine Sci. 3: 421–430.

Beeftink, W.G. 1975. Vegetationskundliche Dauerquadratforschung auf periodisch überschwemmten und eingedeichten Salzböden im Südwesten der Niederlande. In: W.Schmidt, Sukzessionsforschung, Ber. Symp. Intern. Ver. Vegetationskunde 1973, Cramer, Vaduz: 567–578.

Beeftink, W.G. 1979. The structure of salt marsh communities in relation to environmental disturbances. In: R.L. Jefferies & A.J. Davy: Ecological processes in coastal environments. Proc. 1st Eur. Ecol. Symp. Norwich, Blackwell, Oxford: 77–93.

Beeftink, W.G., M.C. Daane, W. de Munck & J. Nieuwenhuize, 1978. Aspects of population dynamics in Halimione portulacoides communities. Vegetatio 36: 31–43.

Böttcher, H. 1975. Stand der Dauerquadrat-Forschung in Mitteleuropa. In: W. Schmidt, Sukzessionsforschung. Ber. Symp. Intern. Ver. Vegetationskunde 1973, Cramer, Vaduz: 31–37.

Braun-Blanquet, J. 1928. Pflanzensoziologie. Grundzüge der Vegetationskunde. Springer, Berlin, 10 + 330 pp.

Clements, F.E. 1916. Plant succession. An analysis of the development of vegetation. Carnegie Inst., Washington, 512 pp.

Connell, J.H. & R.O. Slatyer. 1977. Mechanisms of succession in natural communities and their role in community stability and organization. Amer. Nat. 111: 1119–1144.

Dobben, W.H. van & R.H. Lowe-McConnell (eds.). 1975. Unifying concepts in ecology. Proc. First Internat. Congr. Ecology. Junk, The Hague, 1974, 302 pp.

Grime, J.P. 1978. Interpretation of small-scale patterns in the distribution of plant species in space and time. In: A.H.J. Freijsen & J.W. Woldendorp, Structure and functioning of plant populations. North-Holland, Amsterdam: 101–124.

Harper, J.L. 1977. Population biology of plants. Acad. Press, London, 892 pp.

Hogeweg, P. 1976. Topics in biological pattern analysis. Thesis University of Utrecht, 208 pp.

Horn, H.S. 1976. Succession. In: R.M. May, Theoretical ecology. Principles and applications. Blackwell, Oxford: 187–204.

Kershaw, K.A. 1973. Quantitative and dynamic plant ecology. Arnold, London, 308 pp.

Ketner, P. 1972. Primary production of salt-marsh communities on the island of Terschelling in the Netherlands. Thesis Catholic University of Nijmegen, 181 pp.

Knapp, R. 1974. Handbook of vegetation science. Part VIII. Vegetation dynamics. Junk, The Hague, 364 pp.

Kuhn, T.S. 1970. The structure of scientific revolutions. 2nd Edition. Univ. of Chicago Press, Chicago, 210 pp.

Leeuwen, C.G. van. 1966. A relation theoretical approach to pattern and process in vegetation. Wentia 15: 25–46.

Leeuwen, C.G. van. 1973. Ecologische systeembeschrijving. In: Oecologie. Lecture notes Studium Generale Technical Univ. Eindhoven: 165–185.

Lüdi, W. 1930. Die Methoden der Sukzessionsforschung in der Pflanzensoziologie. In: E. Abderhalden, Handbuch biol. Arbeitsmeth. XI-5 (3): 527–728.

Maarel, E. van der. 1971. Plant species diversity in relation to management. In: A.S. Watt & E. Duffey, The scientific management of plant and animal communities for conservation. Blackwell, Oxford: 45–63.

Maarel, E. van der. 1974. The Working Group for data-processing of the International Society for Plant Geography and Ecology in 1972–1973. Vegetatio 29: 63–67.

Maarel, E. van der. 1975. Man-made natural ecosystems in environmental management and planning. In: W.H. van Dobben & R.H. Lowe-McConnell, Unifying concepts in ecology. Proc. First Internat. Congr. Ecology. Junk, The Hague, 1974: 263–274.

Maarel, E. van der. 1976. On the establishment of plant community boundaries. Ber. Deutsch. Bot. Ges. 89: 415–443.

Maarel, E. van der & P.L. Dauvellier. 1978. Naar een globaal ecologisch model voor de ruimtelijke ontwikkeling van Nederland. I and II. State Authority for Town and Country Planning, Study Rep. 9, 314 + 166 pp.

Maarel, E. van der, L. Orlóci & S. Pignatti. 1976. Data-processing in phytosociology, retrospect and anticipation. Vegetatio 32: 65–72.

MacArthur, R. & E.O. Wilson. 1967. The theory of island biogeography. Princeton University Press, Princeton N.Y., 203 pp.

Margalef, R. 1968. Perspectives in ecological theory. Univ. of Chicago Press, Chicago, 111 pp.

Noordwijk-Puijk, K. van, W.G. Beeftink & P. Hogeweg. 1979. Vegetation development of salt-marsh flats after disappearance of the tidal factor. Vegetatio 39: 1–13.

Odum, E.G. 1969. The strategy of ecosystem development. Science 164: 262–270.

Odum, E.G. 1971. Fundamentals of ecology. 3d Edition. Saunders, Philadelphia, 574 pp.

Orians, G.H. 1975. Diversity, stability and maturity in natural ecosystems. In: W.H. van Dobben & R.H. Lowe-McConnell, Unifying concepts in ecology. Proc. First Internat. Congr. Ecology, The Hague, 1974: 139–150.

Peters, R.H. 1976. Tautology in evolution and ecology. Amer. Nat. 110: 1–12.

Pignatti, S. 1975. Pflanzensoziologie am Scheideweg, Vegetatio 30: 149–152.

Pümpel, B. 1977. Bestandesstruktur, Phytomassevorrat und Produktion verschiedener Pflanzengesellschaften im Glökner Gebiet. Veröff. Oest. MAB-Hochgebirgsprogr. Hohe Tauern 1: 83–101.

Research Institute for Nature Management, Leersum, The Netherlands. Annual Reports 1973–1977.

Schmidt, W. 1974. Bericht über die Arbeitsgruppe für Sukzessionsforschung auf Dauerflächen der Internationalen Vereinigung für Vegetationskunde. Vegetatio 29: 69–73.

Schmidt, W. 1975. Sukzessionsforschung. Ber. Symp. Intern. Ver. Vegetationskunde 1973, Cramer, Vaduz, 622 pp.

Shmida, A. & R.H. Whittaker (in press). Convergent evolution of desert vegetation in the old and new world. Ber. Symp. Intern. Ver. Vegetationskunde 1978, Cramer, Vaduz.

Simberloff, D. 1976. Species turnover and equilibrium island biogeography. Science 194: 572–578.

Steen, W.J. van der. 1973. Inleiding tot de wijsbegeerte van de biologie. Oosthoek, Utrecht, 274 pp.

Strijbosch, H. 1976. Een vergelijkend syntaxonomische en synoecologische studie in de Overasseltse en Hatertse vennen bij Nijmegen, Diss. Nijmegen, Stichting Studentenpers, 335 pp.

Teal, J.M. 1962. Energy flow in the salt marsh ecosystem of Georgia. Ecology 43: 614–624.

Thurston, J.M., E.D. Williams & A.E. Johnston. 1976. Modern developments in an experiment on permanent grassland started in 1856: Effects of fertilizers and lime on botanical composition and crop and soil analyses. Annales Agron. 27: 1043–1082.

Tyler, G. 1971. Distribution and turnover of organic matter and minerals in a shore meadow ecosystem. Studies in the ecology of Baltic sea-shore meadows. IV. Oikos 22: 265–291.

Valiela, I., J.M. Teal & W.J. Sass. 1975. Production and dynamics of salt marsh vegetation and the effects of experimental treatment with sewage sludge. Biomass, production and species composition. J. Appl. Ecol. 12: 973–981.

Watt, A.S. 1960. Population changes in acidiphilous grass-heath in Breckland, 1936–57. J. Ecol. 48: 605–629.

Watt, A.S. 1971. Factors controlling the floristic composition of some plant communities in Breckland. Brit. Ecol. Soc. Symp. No. 11: 137–152.

Werger, M.J.A. & G.A. Ellenbroek. 1978. Leaf size and leaf consistence of a riverine forest formation along a climatic gradient. Oecologia 34: 297–308.

Werger, M.J.A. & W. Schmidt. 1978. Third symposium of the Working Group on succession research on permanent plots. Vegetatio 36: 61–62.

Westhoff, V. 1970. New criteria for nature reserves. New Scient. 46: 108–113.

Westhoff, V. 1976. Die Verarmung der niederländischen Gefässpflanzen flora in den letzen 50 Jahren und ihre teilweise Erhaltung in Naturreservaten. Schriftenreihe für Vegetationskunde 10: 63–73.

Westhoff, V. & E. van der Maarel. 1978. The Braun-Blanquet approach. In: R.H. Whittaker, classification of plant communities. 2nd Edn. Junk, The Hague, p. 287–399.

Williams, E.D. 1978. Botanical composition of the Park Grass Plots at Rothamsted, 1856–1976. Rothamsted Exp. Sta. Harpenden: 1–61.

Whittaker, R.H. 1962. Classification of natural communities. Bot. Rev. 28: 1–239.

Whittaker, R.H. 1975. Communities and ecosystems. 2nd Edition. MacMillan, New York, 385 pp.

MÖGLICHKEITEN ZUR ANWENDUNG VON VEGETATIONSKUNDLICHEN UNTERSUCHUNGEN AUF DAUERFLÄCHEN*

G. LONDO

Research Institute for Nature Management, Leersum, The Netherlands

Keywords:
Dauerflächen, Sukzession, Syndynamik, Permanent quadrats, Succession, Syndynamics

Einleitung

Neben Beschreibungen und Photographien gibt es in der Vegetationskunde zwei andere Methoden zur Untersuchung von Vegetationsänderungen: (a) mittels Dauerflächen-Aufnahmen und (b) mittels periodischer Gebietskartierungen. Ein grosser Vorteil der Kartierung ist, dass man eine viel grössere Oberfläche beschreibt als mit der Dauerflächen-Methode überhaupt möglich ist. Ein Nachteil ist, dass mittels periodischer Kartierungen nur grosse Änderungen registriert werden können. Wenn die zeitliche Änderungen eines Vegetationstypus nicht oder kaum grösser sind als die räumliche Variation innerhalb des Typus, ist die periodische Kartierung weniger geeignet oder man soll längere Zeit mit dem Herstellen der erstfolgenden Karte warten.

Sehr kleine Änderungen in der Vegetation sind mittels Dauerflächen-Untersuchungen wohl gut zu registrieren und hier sind jährliche Beobachtungen, oder sogar mehrere Beobachtungen pro Jahr, sehr sinnvoll. Ein Nachteil aber der Dauerflächen-Methode ist die stark räumliche Beschränkung ihrer Untersuchungsflächen. Diese notwendige Beschränkung zu kleinen Stichproben führt zur Konsequenz, dass innerhalb eines Gebietes nicht alle Sukzessionslinien festgestellt werden können, auch nicht wenn die Zahl der Dauerflächen stark erhöht wird. So wurden zwischen 1963 und 1968 in Dünental-Gesellschaften auf dem Ufer eines ausgehobenen Sees in

den holländischen Dünen mittels 75 Dauerflächen nur 29% der gesamten mittels periodischer Kartierungen herausgearbeiteten Sukzessionslinien entdeckt (Londo 1971, 1975a). Es zeigte sich aus diesen Untersuchungen dass die Kombination beider Methoden den besten Einblick in der Sukzession gab.

Ein Zwischenform zwischen einer 'normale' Kartierung und der Dauerflächen-Methode ist die 'Kartenquadrat-Methode', wobei ein verhältnismässig grosses Quadrat in viele Kleinquadraten zerteilt wird. Diese Kleinquadrate können dann als Gitterwerk für Detail-Kartierungen von Pflanzenpopulationen oder als Grund-Einheiten für Mikro-Aufnahmen dienen. Die Kartenquadrat-Methode ist eine sehr genaue Methode der Vegetationsuntersuchung.

Allgemeine Überlegungen zur Dauerquadrat-Analyse

Falls man sich die räumliche Beschränkung der Flächen realisiert, kann man mit der Dauerflächen-Analyse wertvolle Information über die Sukzession bekommen. Allerdings wären doch einige Punkte zu berücksichtigen:
– zur jährlichen Feststellung einer Sukzession sollen die Aufnahmen verschiedener Jahre vergleichbar sein und dürfen Unterschiede in Saisonaspekten keine Rolle spielen. Bei sehr schnellen Änderungen innerhalb eines Jahres, z.B. in Pioniervegetationen und in benthischen Algen-Gesellschaften, sind mehrere Aufnahmen pro Jahr notwendig.
– Die Genauigkeit der Registrierung hängt von der Feinheit der Skala ab. Weil die Unterschiede in der Zeit im allgemeinen kleiner sind als die räumliche, ist für Dauerflächen eine feine Skala für Deckungsgrad und Abundanz notwendig. Die Dezimalskala (Londo 1974,

* Für die Nomenklatur der Taxas. Heukels-van Oosstroom, 1975. Flora van Nederland. 18e druk. Wolters-Noordhoff, Groningen, 913 pp.; für Syntaxa, s. Westhoff & den Held, 1969. Plantengemeenschappen in Nederland. Thieme, Zutphen, 324 pp.

13

1975a, 1976, Schmidt 1974) ist hierfür sehr geeignet und hat den Vorteil, dass numerische Bearbeitungen damit leicht ausgeführt werden können. Für sehr grosse Flächen ist die Bestimmung des Deckungsgrads aber weniger geeignet und kann man besser die Methode Tansley (1946), mit Buchstabe-Symbolen für Dominanz und Abundanz, anwenden. Für detaillierte Beschreibungen von kleinen Oberflächen ist auch die Punktmethode wertvoll (Braun-Blanquet 1964).

– Oft ist es nicht möglich jedes Jahr alle Dauerflächen zu beschreiben. Man soll dann wenigstens jedes Jahr einen Teil aufnehmen: es sind ja in jedem Jahr Änderungen möglich, z.B. infolge grosser Schwankungen im Niederschlag, die noch viele Jahre nachher in der Vegetationszusammensetzung auswirken und dann oft 'ungreifbar' erscheinen.

– Bei der Feststellung der Oberfläche von Dauerflächen soll man eventuelle Änderungen des Minimumareals der Vegetation während der Sukzession berücksichtigen.

Absichte der Dauerflächen-Analyse

Dauerflächen-Untersuchungen können auf Grund verschiedener Absichten zur Ausführung gebracht werden. Diese Absichte habe ich in drei Kategorien geordnet.

Vegetationskundliche Absichte

Hier handelt es sich wohl vor allem um die Gesetzmässigkeiten der Syndynamik: Aus welchen Gesellschaften kann eine Pflanzengesellschaft entstehen und in welche andere Gesellschaften kann sie übergehen. Hinsichtlich der Syndynamik ist noch vieles unbekannt; die meisten Schlussfolgerungen sind von räumlichen Studien, u.a. von Zonierungsreihen, abgeleitet. Zuverlässige Untersuchungen sind nur diejenige die auf Dauerflächen und periodischer Kartierungen basieren (Beeftink 1977, Londo 1971, Van der Maarel 1975).

Nebst Übergängen in der Zeit von einer Pflanzengesellschaft zu einer anderen, also die Sukzession im engeren Sinne, angehen, sind auch kurzfristige Änderungen innerhalb einer Gesellschaft interessant, z.B. die Fluktuationen infolge Schwankungen im Niederschlag oder im Grundwasserstand.

Wenn die syndynamische Studien durch Untersuchungen der Umweltbedingungen begleitet werden, können die Dauerflächen auch einen wichtigen Beitrag zur Synökologie liefern. Untersuchungen in der Zeit sind vor allem

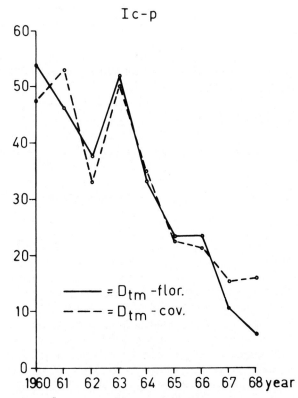

Fig. 1. Änderungen in den durchschnittlichen floristischen (D_{tm}-flor.) und in den durchschnittlichen Deckungs-Änderungsquotienten (D_{tm}-cov.) in der Hygroserie am westlichen Ufer eines ausgehobenen Sees in den holländischen Dünen. In späteren Jahren verursachten grosse Schwankungen im Wasserstand weniger grosse Änderungen in der Vegetation als in früheren Jahren.

Changes in the mean floristic (D_{tm}-flor.) and the mean coverage changeability quotient (D_{tm}-cov.) in the hygrosere on the western shore of an excavated lake in the Dutch dunes. In later years great fluctuations in water level caused less great changes in the vegetation than in former years.

notwendig bei Gesellschaften die an grossen Schwankungen in einem oder in mehreren Umweltfaktoren gebunden sind, z.B. Pioniergesellschaften und viele Gesellschaften des *Agropyro-Rumicion crispi* und des *Nanocyperion*. Scharfe und tiefgehende Kontraste zwischen Faktorenpaaren wie z.B. trocken-nass und salz-süss können für die Keimung und Überlebung verschiedener Pflanzenarten entscheidend sein. Auch können mittels Dauerquadratbeobachtungen und Messungen an Umweltfaktoren viele Ergebnisse hinsichtlich Stabilität und Diversität erhalten werden. So ergab sich in meiner Studie der Ufervegetation einer Dünensee, dass die Vegetation während der Sukzession eine allmählig grössere Anpassung an grossen

14

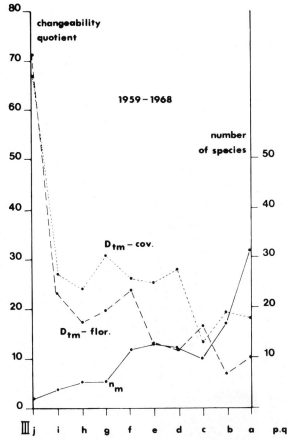

Fig. 2. Die durchschnittliche Anzahl Arten pro Jahr (n_m) und das durchschnittliche floristische und Deckungs-Änderungs-quotient (D_{tm}-flor. und D_{tm}-cov.), berechnet für jede Dauer-fläche des Transekts Nr. III über einer Periode von 10 Jahren
The mean number of species per year (n_m) and the mean floristic and coverage changeability quotient (D_{tm}-flor. and D_{tm}-cov.), calculated for every p.q. of the transect nr. III for a period of 10 years.

Schwankungen im Wasserstand entwickelte (Fig. 1). Die Stabilität in Bezug auf äusseren Faktoren ist also zuge-nommen.

Wenn man den Zusammenhang zwischen räumlichen und zeitlichen Aspekten der Vegetationsentwicklung, also zwischen Struktur und Dynamik, untersuchen will, ist eine gute räumliche Verteilung der Dauerflächen sehr wichtig. Es zeigte sich dass relativ wenige, aber in einem Transekt oder in einigen Transekten ange-ordneten Flächen, viel mehr Information erhalten als viele Flächen willkürlich aufgerichtet. Besonders Dauerflächen welche in Transekten senkrecht auf der Vegetations-zonierung angeordnet sind, verschaffen viele Ergebnisse.

Fig. 2 gibt ein Beispiel aus der Ufervegetation einer Dünen-see. Aus dieser Transektstudie ergab sich übrigens ein negativer Zusammenhang zwischen den Artenreichtum und die relative Vegetationsveränderlichkeit (Londo 1971, 1975b). Besonders für die Studie der Struktur und Dynamik in Pflanzengesellschaften sind Dauertransekte sehr wichtig.

Populations- und autökologische Absichte

In zweiter Linie können die Dauerflächen-Untersuchungen sich auch auf die Autökologie oder die Populations-biologie beziehen: wie verhalten sich die Pflanzenarten an und für sich räumlich und in der Zeit in natürlichen Vegetationen? Auch hierbei ist eine richtige Verteilung der Dauerflächen wichtig. Aus Dauertransekten lassen sich sogenannte Raum-Zeit-Strukturbilder einzelner Arten zusammensetzen. Fig. 3 zeigt die Bilder von zwei Arten in der Ufervegetation einer Dünensee (Londo 1971). *Juncus bufonius* ist ein Beispiel einer Art die in der Zeit leicht von Stelle wechselt; ihre Verbreitung hängt zusam-men mit den Überschwemmung im Winter. Demgegenüber gibt es Arten die in ihrer Verbreitung mehr konstant sind, z.B. *Juncus alpino-articulatus* ssp. *atricapillus*. Im Falle von Mikro-Gradienten sind anliegende zweidimensional ge-ordnete Dauerflächen sehr zweckmässig. Die folgenden Beispiele (Fig. 4) zeigen eine spontane Vegetationsent-wicklung auf einem künstlichen Mikro-Gradienten in meinem experimentellen Garten mit allmählichen Über-gangen zwischen u.a. Sand, Lehm und Ton. *Chrysanthemum leucanthemum*, eine Art von Wiesen und Weiden, hat sich auf alle Substrate ausgebreitet, mit Ausnahme des nähr-stoffarme Sandes. *Rumex obtusifolius*, charakteristisch für Störungsvegetationen, ist aber zurückgegangen und beschränkt sich auf nährstoffreiche Stellen.
Solche Untersuchungen können Ergebnisse über den ökologischen Indikationswert vieler Arten liefern (vgl. Ellenberg 1974).

Absichte der angewandten Pflanzensoziologie

Die dritte Kategorie umfast die angewandte Dauer-flächen-Untersuchungen. Hier können wir wieder zwei Gruppen unterscheiden.

Die erste Gruppe bezieht sich auf die Begründung von Verwaltungsmassnahmen in Naturschutzgebieten. Hierfür ist es notwendig die Einflüsse der verschiedenen Mass-nahmen, wie Mahd und Beweidung, auf die Vegetation kennenzulernen. Daneben gibt es auch Einflüsse aus der Umgebung, z.B. Austrocknung, die auf die Naturschutz-

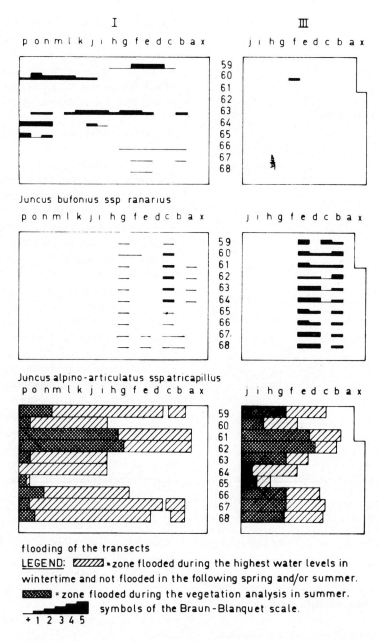

I III

Juncus bufonius ssp ranarius

Juncus alpino-articulatus ssp.atricapillus

flooding of the transects

LEGEND: ▨▨▨ = zone flooded during the highest water levels in wintertime and not flooded in the following spring and/or summer.

▦▦ = zone flooded during the vegetation analysis in summer.

▬ symbols of the Braun-Blanquet scale.

+ 1 2 3 4 5

Fig. 3. Raum-Zeit-Strukturbilder von zwei Arten in Dünental-Gesellschaften im Zusammenhang mit der Überschwemmung.
Space-time patterns of two species in relation to flooding.

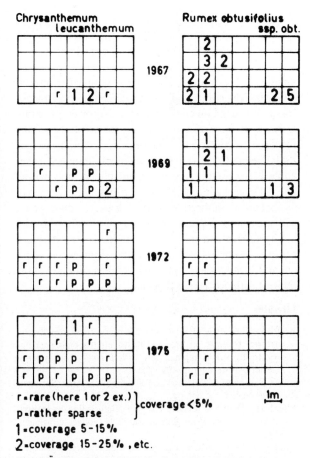

Chrysanthemum leucanthemum   Rumex obtusifolius ssp. obt.

1967
1969
1972
1975

r = rare (here 1 or 2 ex.)
p = rather sparse   } coverage < 5%
1 = coverage 5 - 15%
2 = coverage 15 - 25%, etc.

1m

Fig. 4. Änderungen in der Verbreitung von zwei Arten in einem experimentellen ökologischen Garten (angelegt 1966).
Changes in the pattern of two species in an experimental ecological garden (established in 1966).

gebiete einwirken und sich in der Vegetation ausdrücken.

Die zweite Gruppe von angewandten Dauerflächen-Untersuchungen bezieht sich auf die Nebeneffekte kultur-technischer, ziviltechnischer und industrieller Aktivitäten. Hierzu gehören Untersuchungen von Einflüssen der Luft-verunreinigung, der Infiltration von Flusswasser in den Dünen und Untersuchungen der Brachflächen, Forste, eingedeichte Wattflächen u.s.w. Fig. 5 gibt ein Ergebnis einiger Dauerflächen aus einem Infiltrationsgebiet in den holländischen Dünen, wo eine Vegetationsentwicklung von Gesellschaften mit *Epilobium hirsutum* und *Eupatorium cannabinum* zu einer *Urtica dioica*-Gesellschaft stattfand. Im Laufe der Zeit nahmen sowohl der Artenreichtum als das Differenzquotient ab. Das Differenzquotient ist ein Mass für die äusserliche Verschiedenheit der Dauer-quadrate; je grösser die Unterschiede der Quadrate sind,

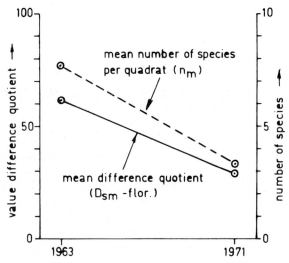

Fig. 5. Abnahme der räumlichen Variation innerhalb einer Reihe von sechs Dauerflächen in einem Flusswasser-Infil-trationsgebiet in den holländischen Dünen (jede Fläche 6m × 0,5m).
Decrease of the spatial variation within a series of six permanent quadrats in a river water infiltration area in the Dutch dunes (each quadrat 6m × 0,5m).

um so grösser das Differenzquotient (Londo 1974, 1975b). Also findet infolge der Infiltration mit sehr eutrophem Wasser aus dem Rhein eine starke Nivellierung in den Dünen statt, wobei die charakteristischen Dünen-Arten verschwinden. In ökologischer Hinsicht ist dieser Einflüss selbstverständlich als negativ zu betrachten (Londo 1975c). Sowohl die angewandten als die rein wissenschaftlichen Dauerflächen-Untersuchungen können beschreibend und experimentell sein.

Auch sind Dauerflächen nicht nur in einer sondern mei-stens in mehreren Hinsichten wichtig. So verschafften meine Dauertransekte in den Dünental-Gesellschaften nicht nur Information über die Sukzession, sondern auch über Struktur und Dynamik und ihren gegenseitigen Zusammenhang und über die Ökologie vieler Arten.

Summary

Besides describing and photographing there are two other methods of registrating vegetation changes: analysis of permanent quadrats (p.q.'s) and successive mapping.
Small changes can only be detected by p.q.'s, but not by successive mapping. A disadvantage of p.q.'s compared with successive mapping is their spatial limitation.
By studying p.q.'s valuable information about vegetation

17

succession can be obtained, provided the following points are taken into account:

– The relevés of the different years must be comparable and differences according to seasonal aspects must be excluded.

– Since differences in time are mostly smaller than those in space, a fine scale is needed for p.q.-analysis. The decimal scale (Londo 1975a) for coverage and abundance is very suitable and has the advantage that calculations are easy. For very large quadrats determination of the coverage is less practical; in those cases the method of Tansley for global estimation of dominance and abundance is more suitable. For detailed studies of small quadrats the pointquadrat method may be useful as well.

– When it is not possible to make a relevé of all p.q.'s every year, it is important to study at least some of them yearly, because extremes in environmental factors can cause changes in the vegetation which remain traceable many years later. These changes may be incomprehensible when the relevé of the extreme year is missing.

– For determining the dimensions of new p.q.'s one has to take into account changes in the minimum area of the plant communities expected during the succession.

Research through permanent quadrats can be done for several purposes.

– The purpose can be concerned with vegetation science itself, notably the syndynamics of plant communities. Most statements about vegetation succession are concluded from spatial studies and little is known which is really based on succession studies. It is clear that the most reliable information can be obtained from analysing p.q.'s and successive mapping.

When also environmental factors are studied, p.q.'s can contribute much to synecology as well. Research-in-time is especially needed and suitable in cases of communities which are bound to great changes in one or more environmental factors. Also information about community stability and diversity can be obtained in this way.

When space-time relationships are to be studied the spatial distribution of the p.q.'s is very important. It became clear that relatively few p.q.'s carefully located in one or some transects gave much more information than many p.q.'s distributed at random. Especially transects across the vegetation zonation gave much information about pattern and process. Thus from a study of dune-slack vegetation a negative correlation could be observed between the species diversity and the relative vegetation changeability.

– P.q.-studies can also be done with respect to autecology and population dynamics: how is the behaviour of plant species in space and time in different vegetation types and in connection with their environmental factors? Permanent transects as well as series of p.q.'s arranged in two dimensions can provide valuable information in this respect.

– The last category is concerned with applied p.q.-studies. Firstly it is done on behalf of the management of nature reserves. It is necessary to study the influence of the management measures on the vegetation types as well as the influences from the environment on the reserve. Further applied p.q.-studies are done to obtain information about the ecological side-effects of human activities a.o. air pollution, infiltration of river water in dune areas, embankment of salt marshes, management of forests, etc.

The applied and the purely scientific studies by p.q.'s can be descriptive or experimental. Studies on p.q.'s are mostly not important for only one but for more of the above mentioned aspects together.

Literatur

Beeftink, W.G. 1977. The coastal salt marshes of western and northern Europe: an ecological and phytosociological approach. In V.J. Chapman (ed.): Wet coastal ecosystems. Elsevier, Amsterdam. p. 109–155.

Braun-Blanquet, J. 1964. Pflanzensoziologie 3. Aufl. Springer, Wien, New York. 865 pp.

Ellenberg, H. 1974. Zeigerwerte der Gefässpflanzen Mitteleuropas. Scripta Geobotanica Band 9. Goltze, Göttingen. 97 pp.

Londo, G. 1971. Pattern and process in dune slack vegetations along an excavated lake in the Kennemer dunes (The Netherlands). Thesis Nijmegen, Verhandeling nr. 2. Research Institute for Nature Management, Leersum, The Netherlands, 279 pp.

Londo, G. 1974. Successive mapping of dune slack vegetation. Vegetatio 29: 51–61.

Londo, G. 1975a. Dezimalskala für die vegetationskundliche Aufnahme von Dauerquadraten. In: W. Schmidt (ed.): Sukzessionsforschung. Cramer, Vaduz. p. 613–617.

Londo, G. 1975b. Information über Struktur, Dynamik und ihr Zusammenhang durch Dauerquadrat-Untersuchungen. In: W. Schmidt (ed.): Sukzessionsforschung. Cramer, Vaduz. p. 89–105.

Londo, G. 1975c. Infiltreren is nivelleren. De Levende Natuur 78(4): 74–79.

Londo, G. 1976. The decimal scale for relevés of permanent quadrats. Vegetatio 33: 61–64.

Maarel, E. van der. 1975. Small-scale changes in a dune grassland complex. In: W. Schmidt (ed.): Sukzessionsforschung. Cramer, Vaduz. p. 123–134.

Schmidt, W. 1974. Bericht über die Arbeitsgruppe für Sukzessionsforschung auf Dauerflächen der Internationalen Vereinigung für Vegetationskunde. Vegetatio 29: 69–73.

Tansley, A.G. 1946. Introduction to plant ecology, 2nd ed. London.

ON PHOTOGRAPHIC TECHNIQUES IN PERMANENT PLOT STUDIES*

D. C. P. THALEN

International Institute for Aerial Survey and Earth Sciences (ITC), 350 Boulevard 1945, 7511 AD Enschede, The Netherlands

Keywords:
Aerial photography, Mapping, Pattern and process, Permanent plots, Remote Sensing, Sampling, Stereo ground photography, Succession.

Introduction – photographic and other imaging techniques

In most European studies of permanent plots sequential photography is used to illustrate field situations, rather than as a tool in data collection. Imaging techniques, however, can be used in the analysis of the patterns and processes under investigation. This communication tries to elaborate on this point. It aims at bridging the gap between two approaches in the study of vegetation succession: detailed vegetation analysis in the field on plots ranging from one to a few hundred square metres and the more general analysis of vegetation pattern changes over larger areas. The role that can be played by photography and photo-interpretation techniques will thereby be discussed and exemplified.

Remote sensing is the term, now well established, to indicate all techniques of data acquisition on a far away area or object. The main value of observations from a distance is to see and·depict objects as a whole or in connection with the whole of which they form a part (Zonneveld 1974). The human eye is by far the best remote sensing system man has available and it is certainly the most used for data collection in vegetation studies. It has the disadvantage, that no lasting image is produced. Photographic techniques are superior in this respect. With the introduction of the camera a comparison of situations recorded at different times became possible. Early this century, Shantz already recognized the applicability of the technique of repeated photography to record changes in the vegetation

(cf. Phillips 1963). With the introduction of photography from the air, starting with single photography from balloon, followed by stereophotography from aircraft and later space photography (Skylab and other missions), new dimensions were added. The introduction of new film/filter combinations has had an enormous impact on the use of photographic techniques in research. The human eye can distinguish only a small number of grey shades as shown on a black and white photograph. The number of colours that can be distinguished is almost unlimited and hence the colourfilm has extended the possibilities for the discrimination of objects.

The electromagnetic spectrum, however, offers more possibilities for remote sensing than only depicting in black and white or colour what the human eye can see. Colour infrared film, often referred to as 'false colour' (a term which encompasses more in its strict meaning), also records reflected energy in the 700–900 nanometer band, beyond the visible part of the spectrum (400–700 nanometer). Because of the high reflection of living vegetation in this part of the spectrum (Woolley 1970), contrary to the reflection by most soil surfaces, it has been found a powerful tool in many vegetation studies, especially where differences in the ratio of bare soil (or water) to vegetation cover exist (e.g. Colwell & Carneggie 1971, Tueller et al. 1972, Seher & Tueller 1973, Driscoll & Coleman 1974, Hayes 1976). It may be emphasized that its usefulness in comparison to panchromatic black and white photography, which is cheaper and much easier to handle, has often been exaggerated (Zonneveld & Thalen 1976).

The term remote sensing, at present encompassing all techniques of sensing from a distance, has for some time often been used exclusively for new techniques like Multi

* Contribution to the Symposium of the Working Group for Succession Research on Permanent Plots, held at Yerseke, The Netherlands, October 1975.

Spectral Scanning from aircraft and satellite, Airborne Thermography and Side Looking Airborne Radar (SLAR). These techniques, although operational and highly useful for particular applications (Hempenius 1976) can for the time being be ignored in permanent plot studies because of the cost involved and/or limitations in resolution. The present satellite remote sensing systems could hardly contribute to succession studies on small areas, several square meters to some hectares. The smallest area that is recorded separately (spatial resolution, pixel (= picture-element)-size) by the scanner on board the LANDSAT-satellites is about 0.45 ha. Moreover, pixels forming the images of consecutive satellite overpasses do not coincide. To be sure that a certain area is depicted as a 'pure' pixel, without any aberration due to adjacent areas with a different reflection, an area considerably larger than 0.45 ha is required. Permanent plots in which the vegetation is recorded in detail over the years are usually of the magnitude of $1-1000$ m^2 and the existing satellite system can therefore be of no direct use here. This does not mean that vegetation science should not carefully watch further developments in these fields (Hempenius 1974).

In stead of taking distance for a better view, as in remote sensing, we could also approach closer in order to see more details separately. In any method of detailed vegetation analysis including permanent plots particular attention should be given to the application of 'proximal imaging techniques'. No sharp boundary can be drawn between 'remote' and 'proximal' sensing. The first gives the object or study area in its environmental setting, in an over-all view, the second gives the detail that can be studied separately. In vegetation succession studies often both are desired and it is the interplay between them, in combination with fieldwork, that should be considered.

Use of aerial photography in establishing permanent plots

In permanent plot studies the decision procedure on 'how many plots of which shape and size are to be placed where', is often superficially treated. Decisions have often been subjective and dictated by preliminary field impressions, expected representativeness, etc. Little can be said against a certain subjective element in this phase of the study, but subjectivity could be reduced by using aerial photographs.

In this first phase of a study a synoptic view of the whole study area is needed. A placing of plots based on a preliminary field inspection alone can lead to an embarrassing non-representativeness of the selected plots, only revealed

at a later stage of the study when the study area becomes better known. Environmental variation and the distribution of the different vegetation types reflecting this variation need to be taken into consideration. Such variation is easily seen on aerial photographs of the proper type and scale. Often vegetation maps have been used. These are abstractions and generalizations of the actual field situation and should be read preferably in combination with aerial photo's.

No details on the most suitable scale and type of photography can be given without a prior knowledge of the study area and study objectives, but a two stage approach including large scale aerial photographic techniques seems most promising.

Good possibilities are offered by a combination of the conventional panchromatic black and white photography at scales of about 1–5000 to 1 : 50000 with sample strips of large scale full colour or colour infrared photography at scales of 1 : 500 to 1 : 5000.

The first type will allow a broad delineation of relatively homogeneous areas for the whole study area, while with the large scale colour imagery the nature or at least the relative homogeneity of the vegetation inside the delineated areas can be detected. This was probably first realized in forest survey sampling (Aldrich et al. 1959) and only later in other vegetation surveys such as those particularly dealing with shrubs (Carneggie & Reppert 1969).

The usefulness of black and white aerial photography for the delineation (not to be confused with identification!) of vegetation types is, as long as photo-scale and study objective reasonably match, beyond discussion after several decades of successful application of the technique.

The possibilities and advantages of large scale high resolution colour and colour infrared photography are less well known, although in detailed forest inventory the technique is already well established (Aldrich 1966). Most commonly used are 70 mm cameras, operated from low (100–200 m) flying light aircraft. Such cameras have fast shutter speeds and rapid film advance, resulting in several images per second. The films are easily processed and the images can be analysed stereoscopically with a pocket stereoscope and the film on the roll (Aldrich 1966, Tueller et al. 1972). Work reported in Carneggie & Reppert (1969) and Driscoll et al. (1970) shows that individual shrub species and even herbaceous vegetation can be identified on colour infrared photography at scales of 1 : 600 to 1 : 1100. Seher & Tueller (1973) designed a key, utilizing height, colour and texture, for the identification of 10 major marsh vegetation types on 70 mm colour infrared aerial photography. A rapid assessment of the herbaceous and shrub communities as desired in establishing permanent plots is now within reach using this technique (Colwell & Carneggie 1971, Driscoll & Coleman 1974). Cost need not be prohibitive, provided a survey aircraft within reasonable distance is available (Tueller et al. 1972, Seher & Tueller 1973).

Ground photography for data collection in permanent plots

Permanent plot studies require a regularly repeated recording of the vegetation inside the plots. This recording should be of a type that allows a comparison in order to work out trends of what is happening to species or, even better, to individual plants. The cover-abundance scale of Braun Blanquet or a refined version of the scale (e.g. Barkman et al. 1964) has often been used for visual estimates. The Working Group for Sucession Research on Permanent Plots has for this purpose adopted the scale proposed by Londo (1976). Density is sometimes worked out by plant unit counts in a subsampling system and cover by a point quadrat method with pins lowered from a frame or along transsects. In these and similar procedures photographic proximal sensing techniques could in many cases speed up data collection, increase accuracy and allow a repetition or the new recording of an attribute in retrospect.

Claveran (1966) used polaroid film and stereoscopy. On the positive prints, immediately available in the field, annotations can be made on the spot and the image quality can be directly assessed. Species can be labelled, particular areas delineated, etc. The stereoscopy considerably decreases problems with overlapping plant overstoreys.

Wimbush et al. (1967) used vertical stereo groundphotography in a succession study on the vegetation of the Snowy Mountains in New South Wales, Australia. Comparing colour transparancies and black and white prints, they found the former giving better results. The prints were analyzed with the usual techniques. To view the transparant 35 mm colour images stereoscopically, the stereopairs were projected with two special mounted projectors from below on to a glass table overlain with white tracing paper. Cover was determined by superimposing a 100-compartment grid on plastic sheeting over the tracing paper. Pierce & Eddleman (1970) reported results from vertical colour stereophotography of 1 sq. m plots, subdivided into 25 equal subquadrats, using artificial light to reduce shadow effects. Comparisons of field information

with photo-interpretation for 15 sample units in vegetation types ranging from medium dense timber stand to open rangeland were made. A 95 per cent accuracy of species identification was possible, most trouble being caused by annual forbs. Later Pierce & Eddleman (1973) reported results on 42 1 sq. m plots photographed when the vegetation was dry. A combination of vertical and oblique stereophotography was used. Cover of several species was underestimated from the photo's as their dry leaves were not always clearly visible. 70 mm photographs were found superior to 35 mm. As a result the use of two cameras operated simultaneously for a vertical as well as low oblique stereopair are recommended. Elimination of shadow is considered most important.

Wells (1971) described a system of vertical ground stereophotography using two cameras mounted in a frame, with a dual shutter release mechanism. The resulting colour transparancies are analysed with a zoom stereoscope applying a hit – technique with cross hairs, in principle similar to a point quadrat method applied in the field.

Instead of using two cameras for the stereo-effect, or one camera operated from two positions, stereophotography may be obtained by readily available stereocameras or a stereo adapter attached to a single lens camera. The last technique is reported in Ratliff & Westfall (1973). For a study in high country in the Sierra Nevada of California where transport was by helicopter and by foot, a rapid field sampling technique had to be used. Vertical stereo groundphotography, with a 35 mm single lens reflex with stereoadapter, taking colour transparancies, worked. With a dot grid overlay and a pocket stereoscope foliar cover and composition could be estimated later behind the desk, provided familiarity with the species had already been gained.

These techniques were used mainly in one-time sampling schemes. They are, however, devised in a way that enables easy repetition of the sampling in time of the same areas. For permanent plot studies information from the papers referred to above, combined with one's personal experiments should easily lead to a photographic recording system most suitable to meet the study objectives and cope with the problems of the vegetation types studied. The usefulness of techniques for photo-monitoring vegetation succession is now increasingly realized (Tueller et al. 1972, McKendrick 1976). An additional advantage of applying repeated photography of plots particularly when management problems are involved is its potential use to illustrate deterioration, improvement, or the effect of certain treatments to those directly involved in the use and management of the land. These persons may not always be familiar with the methods and terminology applied in papers showing research results.

In many vegetation types cover and height, and thereby volume, are highly correlated with aerial biomass and production (Kinsinger & Strickler 1961, Mason & Hutchings 1967, Pasto et al. 1957, Evans & Jones 1958, Uresk et al. 1977, Goebel et at. 1958). The application of photo analysis may, therefore, well be extended to include estimates of these attributes which are of particular importance in many range monitoring studies. A particular case has been reported by Springfield (1974). By repeatedly photographing shrubs against a 1 inch grid production and utilization of the shrubs was estimated by counting the squares partially or totally obscured by plant material.

A final remark may be devoted to sequential photography versus sequential charting. Charting of individual plant units within a permanent plot may be useful but is far more time consuming and at best no more accurate than a photographic method. In case vegetation 'types' within the plot are mapped, the question arises immediately on what basis these types were distinguished. In many cases they will be a generalization of the real situation. In case no photo's are taken, there is no reference material to check the basis for the types. Only a subsampling within the plot could provide such a basis. A quantification of trends based on areas measured on sequential charts can therefore be misleading and is in all cases more time consuming than deriving the same data from proper (annotated) photography.

Aerial photography in extrapolating permanent plot results

When certain vegetation changes within a permanent plot are recorded over a period of time, it remains to be evaluated how representative the recorded changes are for what happened to the vegetation cover over a larger area considered to be represented by the permanent plot. Different approaches can be adopted to allow for such an evaluation, e.g. 1. More than one permanent plot can be established, and followed in time, in what is considered a relatively identical set of environmental conditions. 2. In the course of the study the vegetation inside the permanent plot can be regularly compared with samples of the surrounding area to ensure the continuous representativeness of the permanent plot. 3. Aerial photography of the area concerned may be compared and changes in the photo-characteristics as observed

for the permanent plot compared with changes as seen in the photo-pattern for the rest of the area. Large scale aerial photography (appr. 1 : 1000 to 1 : 5000) would be best for this purpose.

These approaches can be used in combination. Only the repeated photointerpretation technique can show where certain changes took place. Often an originally homogeneous area is broken up into a heterogeneous pattern during the course of the study through an increasing environmental differentiation. In such cases approach (1) will give no and approach (2) very little information on the representativeness of the permanent plot data. A regular coverage of the study area by large scale aerial photography is particularly justified from this point of view. Examples of a 'combined permanent plot – total view' approach may accentuate this.

Results of a vegetation succession study in a dune slack vegetation around an artificial lake in The Netherlands reported in Londo (1974) are of particular interest in this respect. He compared his succession schemes as deduced from vegetation maps with those derived from permanent plots and found that '. . . a much more complete picture of the succession was given by the successive mapping method'. Expressing this in terms of succession lines, Londo found that only 29 per cent of the succession lines derived from successive mappings over a five year period could be obtained from permanent plot analysis. He, therefore, recommended as the most efficient method for a rather complete picture the combination of a restricted number of permanent plots for the elucidation of the year-to-year aspects of the process combined with successive mapping.

Zonneveld (1974) studied vegetation and landscape succession in part of the Biesbosch, a freshwater tidal area in The Netherlands, by photographic monitoring, so far over a period of 25 years. Obliques, taken at least yearly by handheld ordinary camera from a high tension pylon, in combination with verticals at different scales from five years, were available. In addition the author could apply his detailed knowledge of the general dynamics and vegetation processes in the area as obtained during earlier studies. Zonneveld (1974) reports that, under these conditions (1) – geomorphological processes and vegetation (species) succession can be followed in rather great detail in a semi-quantitative way; (2) – the most important species could be recognized individually on the photography; (3) – earlier conclusions on succession derived from field work on side by side occurrence could be largely confirmed.

The value of the photographic monitoring technique is

in this case considerably increased by a detailed knowledge of species, vegetation types and succession, gained by the author during extensive fieldwork. This knowledge is to an extent of a similar type to that gained through detailed work on permanent plots and was in fact partly derived from such studies. The photographic record in combination with detailed work on plots of limited area again seems most valuable.

A last aspect of the application of aerial photography in extrapolating permanent plot results is that aerial photo's if available for certain intervals, can be interpreted at a later date in comparison to each other and in relation to groundtruth collected at approximately the same time. Such work may in retrospect reveal patterns of change in the vegetation. A study of this kind has recently been performed in a coastal saltmarsh and sanddune area at Oostvoorne (Delta area, The Netherlands) by Suij (1977). The area is rapidly changing as a result of human influence, amongst others embanking and introduction of grazing. Combining permanent plot data collected in the field by Beeftink (see Beeftink 1975), new field data, and an interpretation of aerial photograph of 1954, 1961, 1967 and 1972, a series of 1 : 10000 vegetation maps was prepared and succession patterns, both in space and time, could be established.

It is apparently the combination of two types of work, repeated detailed fieldwork in carefully selected small areas, and repeated study of larger areas to extrapolate the permanent plot results, that gives the best overall results in following succession over larger areas. Photographic techniques can be a valuable tool in both. For the second type they may be regarded an indispensable tool, particularly when proper management of larger areas is a study objective.

Summary and conclusions

In all studies where permanent plots are used as a sampling method for vegetation processes over larger areas, the representativeness of the processes recorded within the plots for what happened to the vegetation of the area as a whole should be given due consideration. Aerial photography of a type to be selected for each study area separately has proven to be a highly efficient and in many cases indispensable tool for this purpose.

Repeated stereo colour groundphotography can be a useful method in permanent plot studies. It allows detailed measurements on important vegetation attributes and can reduce (sometimes costly) fieldwork. Raw field data can be easily stored and measurements repeated or taken at a later date. Trends can be illustrated easily.

To enable the extrapolation of permanent plot results to a wider area and the detection of spatial patterns of change, repeated aerial photography of not too small a scale is the most efficient method. It is more reliable than repeated mapping in the field because of generalization and typification problems inherent to the later method.

For a succession study in a not too limited area a combination of conventional aerial photography (black and white panchromatic vertical stereo at scales 1 : 5000 to 1 : 50000), large scale 70 mm photography (stereo colour or colour infrared at scales 1 : 500 to 1 : 5000) and groundwork in permanent plots, including ground photography seems the most promising approach. In rangeland monitoring the idea is not new but such an integrated approach deserves wider application.

References

Aldrich, R.C. 1966. Forestry applications of 70 mm color. Photogram. Eng. 32: 802–810.

Aldrich, R.C., W.E. Bailey & R.C. Heller. 1959. Large scale 70 mm color photography techniques and equipment and their application to a forest sampling problem. Photogram. Eng. 25: 747–754.

Almkvist, B. 1975. The influence of flight altitude and type of film on photo interpretation of aquatic macrophytes. Svensk Bot. Tidskr. 69: 181–187.

Barkman, J.J., H. Doing & S. Segal. 1964. Kritische Bemerkungen und Vorschläge zur quantitativen Vegetationsanalyse. Acta Bot. Neerl. 13: 394–419.

Beeftink, W.G. 1975. Vegetationskundliche Dauerquadratforschung auf periodisch überschwemmten und eingedeichten Salzböden im Südwesten der Niederlande. In: W. Schmidt (Ed.), Sukzessionsforschung. Ber. Symp. Intern. Ver. Vegetationskunde 1973, Rinteln, 1975: 567–578.

Carneggie, D.M. & J.N. Reppert. 1969. Large scale 70 mm aerial color photography. Photogram. Eng. 35: 249–257.

Claveran, R.A. 1966. Two modifications to the vegetation photographic charting method. J. Range Manage. 19: 371–373.

Colwell, R.N. & D.M. Carneggie. 1971. Applications of remote sensing to arid-lands problems. pp. 173–186 in: Food, fiber and the arid lands (W.G. McGinnies, B.J. Goldman & P. Paylore, eds.). The Univ. of Arizona Press, Tucson.

Driscoll, R.S., J.N. Reppert & D.M. Carneggie. 1970. Identification and measurement of herbland and shrubland vegetation from large scale aerial colour photographs. Proc. XI Int. Grassl. Congress, pp. 95–98.

Driscoll, R.S. & M.D. Coleman. 1974. Color for shrubs. Photogram. Eng. 40: 451–459.

Evans, R.A. & M.B. Jones. 1958. Plant height times ground cover versus clipped samples for estimating forage production. Agron. J. 50: 504–506.

Goebel, C.J., L. Debano & R.D. Lloyd. 1958. A new method of determining forage cover and production on desert shrub vegetation. J. Range Manage. 11: 244–246.

Hayes, F. 1976. Application of color infrared 70 mm photography for assessing grazing impacts on stream-meadow ecosystems. Station note no. 25 Univ. of Idaho, Forest, Wildlife, and Range Exp. Sta. pp. 3.

Hempenius, S.A. 1974. How can ecology prepare itself for remote sensing? Proc. First Int. Congress of Ecology, Junk b.v.. Publ. The Hague, 1974. (also in: ITC-Journal 1974 (4): 561–571).

Hempenius, S.A. 1976. Critical review of the status of remote sensing. Bildmessung und Luftbildwesen 44: 29–41.

Kinsinger, F.E. & G.S. Strickler. 1961. Correlation of production with growth and ground cover of Whitesage. J. Range Manage. 14: 274–278.

Londo, G. 1974. Successive mapping of dune slack vegetation. Vegetatio 29: 51–61.

Londo, G. 1976. The decimal scale for relevés of permanent quadrats. Vegetatio 33: 61–64.

Mason, L.R. & S.S. Hutchings. 1967. Estimating foliage yields on Utah Juniper from measurements of crown diameter. J. Range Manag. 20: 161–166.

McKendrick, J.D. 1976. Photo-plots reveal arctic secrets. Agroborealis Jan. 1976: 25–30.

Pasto, J.K., J.R. Allison & J.B. Washko. 1957. Ground cover and height of sward as a means of estimating pasture production. Agron. J. 49: 407–409.

Phillips, W.S. 1963. Photographic documentation vegetational changes in Northern great plains. Univ. of Arizona Agr. Exp. Sta. report no. 214, pp. 185.

Pierce, W.R. & L.E. Eddleman. 1970. A field stereophotographic technique for range vegetation analysis. J. Range Manage. 23: 218–220.

Pierce, W.R. & L.E. Eddleman. 1973. A test of stereophotographic sampling in grasslands. J. Range Manage. 26: 148–150.

Ratliff, R.D. & S.E. Westfall. 1973. A simple stereophotographic technique for analyzing small plots. J. Range Manage. 26: 147–148.

Seher, J.S. & P.T. Tueller. 1973. Color aerial photo's for marshland. Photogram. Eng. 39: 489–499.

Springfield, H.W. 1974. Using a grid to estimate production and utilization of shrubs. J. Range Manage. 27: 76–78.

Suij, M. 1977. Veranderingen in de vegetatie van het groene strand van Oostvoorne onder invloed van de afsluiting van het Brielse Gat en van beweiding door koeien, paarden en geiten. Report no. D4-1977, Delta Institute for Hydrobiological Research, Yerseke, The Netherlands.

Tueller, P.T., G. Lorain, K. Kipling & C. Wilkie. 1972. Methods for measuring vegetation changes on Nevada rangelands. Report no. T16 Agric. Exp. Sta. Univ. of Nevada, Reno, Nevada, pp. 55.

Wells, K.F. 1971. Measuring vegetation changes on fixed quadrats by vertical ground stereophotography. J. Range Manage. 24: 233–236.

Wimbush, D.J., M.D. Barrow & A.B. Costin. 1967. Color stereophotography for the measurement of vegetation. Ecology 48: 150–152.

Woolley, J.T. 1971. Reflectance and transmittance of light by leaves. Plant Physiol. 47: 656–662.

Uresk, D.W., R.O. Gilbert & W.H. Richard. 1977. Sampling big sagebrush for phytomass. J. Range Manage. 30: 311–314.

Zonneveld, I.S. 1974. 25 years of sequential photographic monitoring of a tidal environment. Proc. Symp. Comm. VII I.S.P., 7–11 Oct. 1974 Banff, Alberta, Canada (also in: ITC-Journal 1974 (3): 377–384).

Zonneveld, I.S. 1974b. Aerial photography, remote sensing and ecology. Proc. First Int. Congress of Ecology, Junk b.v. Publ., The Hague, 1974. (also in: ITC-Journal 1974 (4): 553–560).

Zonneveld, I.S. & D.C.P. Thalen. 1976. Methods in vegetation survey for development – the present state of the art. Workshop – paper Int. Symp. 'Surveys for Development' 9–15 Dec. 1976 ITC, Enschede, The Netherlands, 20 pp.

UPROOTED TREES, THEIR DISTRIBUTION AND INFLUENCE IN THE PRIMEVAL FOREST BIOTOPE[*][**]

Janusz Bogdan FALINSKI

Geobotanical Station, Bialowieza, Poland

Keywords:
Pino-Quercetum, Poland, Primeval forest, Rate of decay, Succession, Tilio-Carpinetum, Vegetation pattern, Windbreak, Windfall.

Introduction

It is commonly accepted that standing snags, and wood in different states of decay on the ground, are a characteristic element in the primeval forest landscape (Paczoski 1928, Rubner 1918, Fröhlich 1954, Leidengut 1959, Košir 1970, Faliński 1975). Continual accumulation of decaying wood and the process of decay itself are important links in the uninterrupted cycle of energy and matter transformation. It is also accepted that uprooted trees can play a relevant role in the forest ecosystem. Most prominently, they affect the internal structure and dynamics of the ecosystem resulting in changes in the vertical and horizontal structure. This has far reaching consequences in the amount of radiation energy received which change the microclimate and modify microbial and other biological processes. In the decaying wood on the forest floor successions initiate (cf. Hackiewicz-Dubowska 1936) as well as on the mineral soil exposed by the uprooted trees, especially in the case of species, such as *Picea abies*, with extensive root plates. Through these, a thorough differentiation and continuous transformation of the forest environment occur.

The present paper gives results from investigations on permanent study sites in the Bialowieza National Park, Poland, which were carried out by the author and his associates between 1964–1975. The immediate circumstances that result in uprooting trees, the distribution of windfall and windbreak over the forest floor, the time

factor involved, and the subsequent differentiation and transformation of the forest biotope are discussed.

Objectives and Methods

The Bialowieza National Park preserves the fragments of the Lowland Forest with virgin stands of natural origin. The park area, covering about 47 km², was established more than half a century ago. The stands are multilayered and they are composed of different species. They are characterized by a broad age distribution, with many of the trees being more than 200 years old, and some even older than 400 years. Specimens of most species reach exceptionally large sizes. It is not unusual to find individuals of *Picea abies*, for instance, higher than 54 m, and *Pinus sylvestris*, *Quercus robur*, *Tilia cordata*, or *Fraxinus excelsior* up to 40–42 m. Individuals of *Picea abies* may have trunks with diameters up to 140 cm, *Pinus sylvestris* up to 170 cm, *Quercus robur* up to 230 cm, *Fraxinus excelsior* and *Tilia cordata* up to 200 cm, and *Populus tremula*, *Alnus glutinosa*, *Betula verrucosa* up to 100 cm as reported by Zareba (1968). The trees can reach particularly large dimensions in the most productive habitats of the Bialowieza Primeval Forest which are a part of the hygrophilous *Tilio-Carpinetum*.

Conditions under which trees fall and the spatial distribution of these, were analysed in the forest communities of the *Tilio-Carpinetum* (Fig. 1) and *Pino-Quercetum* within two permanent study sites. The windfall and windbreak occurring in the autumn of 1971 in the *Tilio-Carpinetum* and those between the autumn of 1971 and the spring of 1972 in the *Pino-Quercetum* are described in

---

[*] Nomenclature of species follows Flora Europaea.
[**] Contribution to the Symposium of the Working Group for Succession Research on Permanent Plots, held at Yerseke, the Netherlands, October 1975.

25

Fig. 1. Windbreak in *Tilio-Carpinetum*. **The** photograph was taken in the autumn, two years after the destruction took place (photo J. Herezniak).

Table 2. Each tree trunk on the ground received an inventory number on two metal plates. One was tied around the stem by a nylon thread and the other was placed on a metal bar stuck in the ground at the base of the trunk. The rate of increase and decrease in the number of uprooted trees was analysed on a 1 ha area in the *Tilio-Carpinetum* in the period 1964–1975 (Figs. 2, 3 and 7). The method consisted mainly of mapping newly fallen trees and their disappearance as a result of decomposition.

Results

Investigations on the conditions under which the trees fell and their position in the forest were based on 250 trees (Table 1).

The species compositions of the two samples are identical to the floristic composition of the tree layer in the communities, except *Picea abies* whose proportion is higher in the windfall than in the stand.

It was found that, in general, at the time of falling 45 per cent of the trees were already dead (55 per cent alive). In *Picea abies*, an almost equal proportion of dead and living trees were affected. In 87 per cent of the cases, the

26

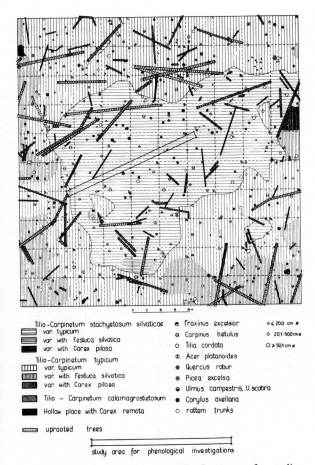

Tilio-Carpinetum stachyetosum silvaticae
- var. typicum
- var. with Festuca silvatica
- var. with Carex pilosa

Tilio-Carpinetum typicum
- var. typicum
- var. with Festuca silvatica
- var. with Carex pilosa

Tilio - Carpinetum calamagrostietosum

Hollow place with Carex remota

uprooted trees

⊖ Fraxinus excelsior   ○ < 20,0 cm ∅
⊖ Carpinus betulus   ○ 20,1-50,0 cm ∅
○ Tilio cordata   ○ > 50,1 cm ∅
⊕ Acer platanoides
⊕ Quercus robur
● Picea excelsa
⊖ Ulmus campestris, U. scabra
● Corylus avellana
○ rotten trunks

study area for phenological investigations

Fig. 2. The study area (1 ha) in *Tilio-Carpinetum* for studies within IBP. Shown are the locations of uprooted trees and the distribution of standing trees within compartments based on vegetation types in 1964.

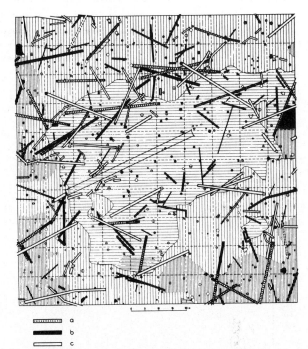

a
b
c

Fig. 3. Changes in the state of uprooted trees form 1964 to 1974 (*Tilio-Carpinetum*): (a) Uprooted trees occurring before 1964 and still existing in 1974, (b) Uprooted trees occurring in 1964 and decomposed before 1974, (c) Uprooted trees occurring after 1964 and still existing in 1974.

immediate cause of falling was uprooting by wind and in 12 per cent crashing by falling trees. The data for the particular species and study areas are given in Table 2. Among the fallen trees all age and height classes are represented (from 10 to 50 m height) with a marked dominance of the age group of 40–80 years with 21–30 m heights between (Fig. 4).

Contrary to the impression one gets at entering the forest, the distribution of windfall forms a recognized regular pattern. That *Picea abies* falls in the direction of the prevailing wind can be seen from the diagram: in Figs. 5 and 6 the centers represent the bases of the trees and the radial line segments depict their position on the ground. The lengths of the line segments are proportional to the trunk lengths. Trees uprooted in autumn, in the period of prevailing western and north-western winds, mostly are oriented in a south-eastern or eastern direction, as is evident in the case of *Picea abies* in the *Tilio-Carpinetum* (Fig. 5). The joint effect of autumn and spring winds, coming from opposite directions in the *Pino-Quercetum* is shown in Fig. 6. The situation is

Table 1. Number of fallen trees in the study sites.

Tilio-Carpinetum		Pino-Quercetum	
Picea abies	89	Picea abies	74
Carpinus betulus	16	Pinus sylvestris	19
Tilia cordata	5	Betula verrucosa	18
Acer platanoides	2	Quercus robur	12
Quercus robur	1	Populus tremula	4
Total	113		137

Tab. 2. Circumstances of windfall, distribution and location of dead and living trees at the moment of falling

FOREST COMMUNITIES Treespecies	n	State of tree at the moment of falling		Immediate circumstances of falling of tree			Location on forest ground			
		living	dead	wind	falling of other trees	other	Lying on the ground	hanging on other trees	propped against standing trees	other
TILIO-CARPINETUM	113									
Picea abies	89	49	51	94	5	1	20	4	58	18
broad-leaved	24	100	0	75	25	0	54	0	17	29
PINO-QUERCETUM	137									
Picea abies	79	49	51	86	14	0	42	10	35	13
Pinus sylvestris	19	32	68	89	11	0	58	5	32	5
broad-leaved	39	95	5	79	18	3	51	10	3	36

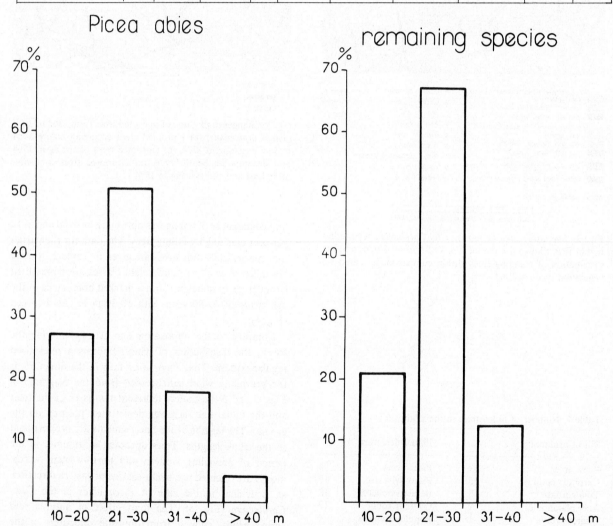

Fig. 4. Distribution of height classes of trees recorded as freshly uprooted trees in the autumn in the *Tilio-Carpinetum*.

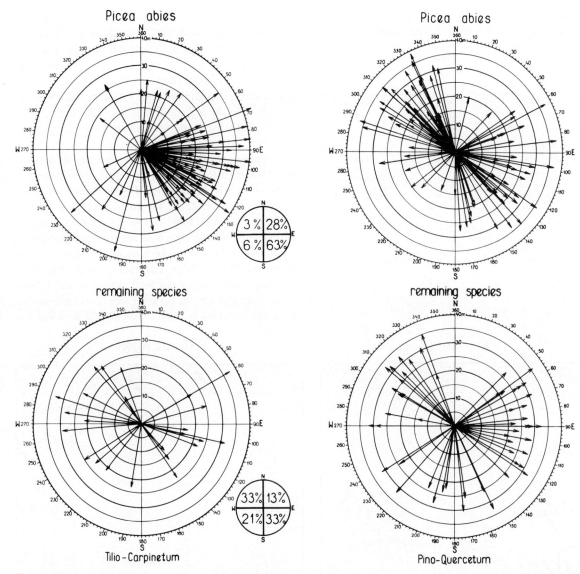

Fig. 5. Direction in which uprooted trees fall in the *Tilio-Carpinetum*. The orientation of the *Picea abies* logs on the ground indicate the direction of the impact of western and north-western autumn winds. The length of the arrows indicates the length of the trunks.

different in both forest communities as regards other tree species: these frequently lie oriented according to wind direction but sometimes they occur in a different position, probably not because of the direct impact of the wind, but air turbulence. Some trees may have been knocked down or were thrown out of balance by earth movement in the neighbourhood of large uprooted *Picea* trees.

Fig. 6. Direction in which uprooted trees fell from autumn 1971 to spring 1972 in the *Pino-Quercetum*. The directions of the logs on the ground indicate wind directions in the autumn and spring. The lengths of the arrows indicate the lengths of the trunks.

From investigations on large surface areas it is known that about 200 to 450 tree trunks per 100 ha per year accumulate on the forest floor. Trees are uprooted all over the year, but show maxima in autumn and spring, owing to strong winds in those seasons. In a 1 ha area of the *Tilio-Carpinetum* uprooted trees, recorded and mapped in 1964 (Fig. 2, Tab. 3), had a joint stem length of 1125 m and a total volume of 60 m³ of wood in bark without

29

Tab. 3. Balance of uprooted trees on a 1 ha study area in Tilio-Carpinetum for the period 1964-1974

	Total stem length	Stem volume
State in 1964	1125 m	60.0 m^3
Decrease 1964-1974	563 m	29.0 m^3
Remaining from 1964	562 m	31.0 m^3
Increase 1964-1974	739 m	40.3 m^3
State in 1974	1301 m	71.3 m^3

branches. In the following years (Fig. 3, Tab. 3) 8 to 13 new trees with a total volume of about 40 m^3 and an individual length of about 13 to 21 m were added annually. During the last year of the study a drastic increase in the number of newly fallen trees occurred. The yearly decrease per ha owing to decomposition of the logs was 2.9 m^3 on average. Thus, both processes differed widely but the short period of observation does not allow to suggest any trend (Fig. 7).

Comparison of the values for increase and decrease of wood volume in logs owing to accumulation and decomposition of the tree stems during the 10-year period allows, however, an estimate of the balance of dead organic matter in the forest (Table 3). This balance for the 10-year period indicates that the decrease by decomposition is compensated by continuous new accumulation of tree trunks on the forest floor. The latter, however, was about 1.6 times higher than the annual increase in the volume of the trees standing on the same area. This suggests that the examined stand is in a phase of natural thinning. The

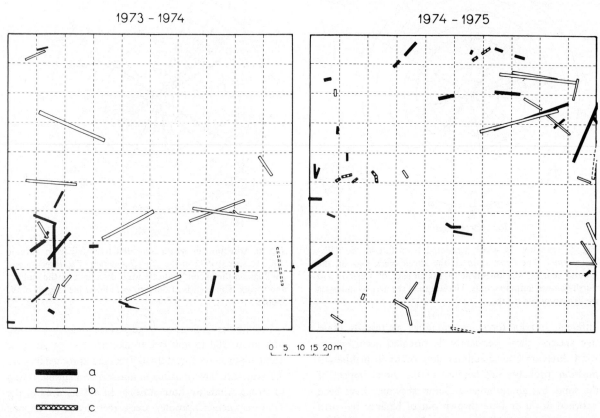

1973 - 1974   1974 - 1975

0 5 10 15 20m

a
b
c

Fig. 7. Example of an annual cycle of changes in the state of uprooted trees in the *Tilio-Carpinetum*: (a) Logs composed, (b) New windfall, logs persisting, (c) New windfall, logs decomposed.

30

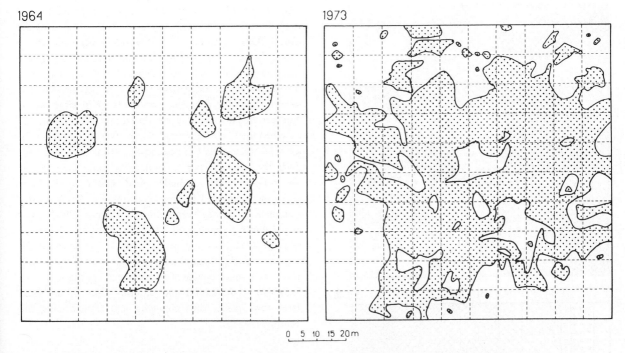

1964   1973

0 5 10 15 20m

Fig. 8. Changes in the abundance of heliophilous herbs (mainly *Urtica dioica*) in the *Tilio-Carpinetum* as a consequence of natural thinning during 1964–1974.

changes in the herb layer manifested by the dominance of heliophilous herbs (mainly *Urtica dioica*) in the summer period is characteristic (Fig. 8).

The development of a forest phytocenosis up to a phase of complete degeneration as described picturesquely and documented for a fragment of *Abies-Fagus* forest in the Alps by Zukrigl et al. (1963; Walter 1973), may be observed rather frequently in the Bialowieza National Park. All the known cases, however, involve conifers (*Peucedano-Pinetum, Sphagno girgensochnii-Piceetum*) or mixed coniferous stands composed of *Picea* and *Quercus* (*Pino-Quercetum, Querco-Piceetum*). In the mixed stånds of the *Tilio-Carpinetum* the broad age distribution would prevent the phytocenosis to reach a phase of total degeneration. This type of forest seems to go through phases of· dominance with different tree species such as a *Tilia, Carpinus, Quercus* and *Picea* phase. In the Bialowieza Forest we find numerous examples of this. Recently they have been documented by Wloczewski (1972) in investigations on permanent areas of 34 years duration.

Decaying wood and craters left by root plates constitute a special type of microenvironment. Their density and localization change continuously owing to the occurrence of new uprooted trees and the disappearance of old ones.

The craters left by *Picea* roots are 0.3–0.7 m deep and have surface areas of 2 to 40 m². In a 1 ha area of *Tilio-Carpinetum*, 30 such craters were found simultaneously. .The continual formation of new craters under newly uprooted trees importantly differentiate the microrelief and the moisture conditions in the forest biotope. Their role as a geomorphological factor has been described by Brzozowski (1963). Their importance as a factor in soil development in the Bialowieza Forest has been described by Prusinkiewicz & Kowalkowski (1964).

The craters are often locals of non-forest species, such as aquatic, silt and mud therophytes. These species constitute for a long time an extrinsic element embedded in the ecosystems, but given time slowly disappear due to competition. Initially, this is prevented by water accumulating in the depressions and by animal activity. In broad-leaved and mixed forests, the craters are subject to plant succession beginning with a phase of aquatic and bog plants. Synusiae with *Carex remota* and *Rumex sanguineus* frequently are the early invaders in the *Tilio-Carpinetum*. Silting up of the craters leads to the disappearance of *Carex remota* and other companion species from the site in about 8 years.

The investigations in the permanent sites have shown

31

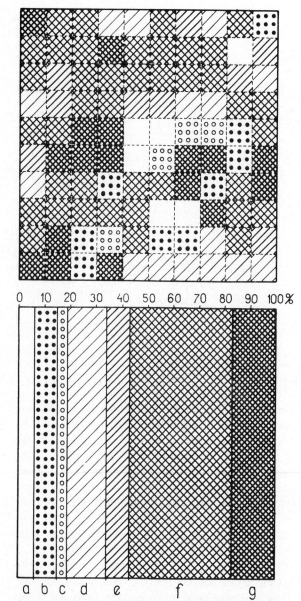

Fig. 9. Differentiation of the site in the *Tilio Carpinetum* regarding changes in uprooting by windfall during 1964–1974: a) plots without windfall (uprooted trees), b) plots without windfall in 1964 but with some in 1974, c) plots with windfall in 1964 and without in 1974, d) plots with the same windfall in 1964 and in 1974, but partly decomposed, e) plots with the same windfall in 1964 as in 1974, but with some new accumulation, f) plots in which some of the windfall completely decayed during 1964–1974 and some new accumulated, g) plots on which the old windfall was completely replaced during 1964–1974.

that in the *Tilio-Carpinetum* a large enough number of trees are uprooted yearly to cover 12–15 per cent of the forest floor. Considering that about 4 per cent of the surface area is occupied by standing trees and 2 per cent covered by craters, the total mineral surface inaccessible to species in the understorey is 17–20 per cent. In the mixed forest the decaying wood collects in large heaps, but scattered distribution over the entire surface can occur.

When the 1 ha area is divided into 100 1 are units (Fig. 9), only six such units were found free from uprooted trees over the whole 10-year period. Nine units were free at the beginning but gradually filled with windfall. Four units were occupied by uprooted trees in 1964 but after complete decay they were free in 1974. A partially decayed wood cover characterized 17 units. A total exchange of old uprooted trees and new arrivals in the course of 10 years occurred in 20 units and a partial one in 44 units. This means that in the course of 10 years in 2/3 of the area, very intensive accumulation and decomposition of uprooted trees took place, thus changing and transforming the biotope (Fig. 9).

Conclusions

It is obvious that ecological consequences of windfall and decomposition are significant for a larger surface area than that of the area disturbed directly by uprooting or under decaying wood. In time, the ecological effects are cumulative and very consignential. The saprophytes, such as *Oxalis acetosella*, *Geranium robertianum* and *Phegopteris dryopteris*, are a case in points. These colonize the sites where decaying wood accumulates. Among cryptogams, soil organisms and some other annuals, one will undoubtedly find indicators of the end stage of wood decay. Fourteen years of observations on phenology in a related study by the author have shown that the sites of decomposed tree trunks, even after total decay and disappearance, are inaccessible to many species, such as for instance *Allium ursinum*, for 5–10 years.

The transformation of the forest biotope due to accumulation of decaying wood, is therefore, a factor contributing to the formation of a great number of different ecological niches inhabited by specialized plant groups, animals and microorganisms. Scores of examples have been described in the Bialowieza Forest as well as in other forest complexes, to show this, but so far they cannot be pieced together into one picture and one ecological process. It is worth mentioning that the studies in the Bialowieza National Park revealed that logs on the ground form the preferred pathways of forest rodents, particularly during the night (Olszewski 1963, 1968). During studies of the effect of wild boar, on the structure and dynamics of the *Tilio-Carpinetum* (Falinski, unpublished), it was noted that accumulation of uprooted trees can form an obstacle to grouting.

The importance of windfall and of the root craters from wind fell trees, as differentiating and transforming factors, may be compared only with the role of large hoofed animals such as the bison and the wild boar. This is particularly true in forests on flat terrain where there is no water or wind erosion which would modify the forest biotopes and biocenoses.

A more detailed description and explanation of changes brought about in the forest biotope by uprooted trees is possible with the help of pedologic research (Brown 1976).

Summary

The paper reports results of a long-term (1964–1974) investigation on permanent study sites in natural forest ecosystems of the *Tilio-Carpinetum* and the *Pino-Quercetum* in the Bialowieza Forest. The influence of decaying logs and root craters was investigated. It was found that the main causes of uprooting were the spring and autumn winds. Wind direction and the position of logs lying on the ground are correlated. *Picea* is most susceptible to uprooting by winds. Almost one half of the trees of this species are alive at the moment of uprooting.

By mapping changes in the distribution of uprooted trees on a permanent area in time, a balance of the change over in a 10-year period was determined. It appeared that the decomposition is slower than accumulation. From this, it was concluded that the stand is in a phase of natural thinning. In the study site, compartments were disinguished with various degrees of change in the number of uprooted trees, and the consequences of differentiation and constant transformation of the biotope and biocenosis by the occurrence of uprooted trees and by their decay are described.

References

Brown, J.L. 1976. Etude de la perturbation des horizons du sol par un arbre qui se renverse et de son impact sur la pédogénèse. Can. J. Soil Sci. 57: 173–186.

Brzozowski, S. 1966. Mapa typologiczna doliny Lopusznej. In: Kondracki, J. (ed.), Sympozjum w sprawie regionalizacji fizyczno-geograficznej Polski i krajów osciennych. Przewodnik wycieczki. Pol. Tow. Geogr., Warszawa.

Falinski, J.B. 1977. In: Falinski, J.B. & J. M. Herezniak. Zielone grdy i czarne bory Bialowiezy. Inst. Wyd. Nasza Ksigarnio, Warszawa.

Fröhlich, I. 1954. Urwaldpraxis. Neuman, Berlin.

Hackiewicz – Dubowska, M. 1936. Roślinność gnijacych pni w Puszczy Bialowieskiej. Sprawozd. z. pos. Jow. Nauk. Warszawa 29 (4): 189–222.

Košir, Z. 1970. Beitrag zur Erforschung der Urwaldstruktur reiner Buchenwälder, In: R. Tüxen (ed.), Gesellschaftsmorphologie. Ber. Int. Symp. Rinteln 1966. pp. 306–314. Junk, The Hague.

Leidengut, H. 1959. Über Zweck und Methodik der Struktur- und Zuwachsanalyse von Urwaldern. Schweiz. Zeitschr. f. Forstw. 3: 111–124.

Olszewski, J.L. 1963. Wechsel der Bewegung der Nager im Walde. Acta Theriologica 7: 372–373.

Olszewski, J.L. 1968. Role of uprooted trees in the movements of rodents in forest. Oikos 19: 99–104.

Paczoski, J. 1928. La végétation de la Forêt de Bialowieza. V-ne Exc. Phytogéogr. Intern. Edit. d. Minist. de l'Agricult., série E, 87 pp. Varsovie.

Prusinkiewicz, Z. & A. Kowalkowski. 1964. Studia gleboznawcze w Bialowieskim Parku Narodowym. Roczn. Glebozn. 14 (2): 161–304.

Rubner, K. 1928. Urwald oder Kulturwald? Bialowies in deutscher Verwaltung 4: 273–285. Paul Parey, Berlin.

Walter, H. 1973. Algemeine Geobotanik. Eugen Ulmer, Stuttgart.

Wloczewski, T. 1972. Dynamika rozwoju drzewostanow w oddziale 319 Bialowieskiego Parku Narodowego. Folia Forest. Pol. A, 20: 5–37.

Zarba, R. 1968. Drzewa i drzewostany BPN. In: Falinski, J.B. (ed.), Park Narodowy w Puszczy Bialowieskiej. pp. 228–236. PWRil, Warszawa.

Zukrigl, K., G. Eckhardt, & J. Nather. 1963. Standortskundliche und waldbauliche Untersuchungen in Urwaldresten der niederösterreichischen Kalkalpen. Mitt. Forst. Bundes. Versuchsanstalt Mariabrun 62: 1–244.

EVOLUTION D'UNE GARRIGUE DE QUERCUS COCCIFERA L. SOUMISE A DIVERS TRAITEMENTS: QUELQUES RESULTATS DES CINQ PREMIERES ANNEES*·**

P. POISSONET, F. ROMANE, M. THIAULT & L. TRABAUD

Département d'Ecologie générale, Centre d'Etudes Phytosociologiques et Ecologiques Louis Emberger (C.E.P.E.), C.N.R.S., Route de Mende, B. P. 5051, 34033 Montpellier Cedex, France

Keywords:
Evolution, Expériences, Garrigue

Introduction

Le terme de 'garrigue' désigne, dans la région méditerranéenne française, plusieurs formations végétales (au sens de Godron et al. 1968) dont les plus fréquentes sont:
- des formations ligneuses hautes (supérieutes à 2 m),
- des formations ligneuses basses (inférieures à 2 m),
- des formations complexes.

La garrigue qui couvre environ 400 000 hectares résulte de l'action très ancienne de l'homme (plus de 3 000 ans). Les éléments principaux de cette action ont été le feu, le pâturage, la coupe des ligneux hauts et la mise en culture.

C'est pour tenter de mieux comprendre l'évolution de cet écosystème et de comparer les différentes formes d'action de l'homme que des expériences sont faites, depuis 1969, dans une garrigue de Chêne kermès (*Quercus coccifera*) qui est une formation ligneuse basse. Quelques observations sur l'évolution du nombre d'espèces et sur le comportement de quelques espèces sont présentées dans ce texte.

Cette étude de la garrigue intervient au moment où de profondes modifications apparaissent dans l'utilisation de ce milieu. Ainsi, l'élevage disparaît progressivement depuis plusieurs décennies; le taillis de Chêne vert (*Quercus ilex*) n'est plus utilisé pour le bois de chauffage; une

---

* Nomenclature d'après 'Flora Europaea' (Tutin et al., 1964, 1968, 1972, 1976) ou 'Les Quatre Flores de France' (Fournier, 1961) pour les espèces non encore décrites dans 'Flora Europaea'.

** Contribution to the Symposium of the Working Group for Succession Research on Permanent Plots, held at Yerseke, The Netherlands, October 1975.

urbanisation de plus en plus grande est en train de se développer.

Les expériences du Puech du Mas du Juge

Le choix de la station

Les expériences ont été implantées à 10 km au Nord de Montpellier, sur le versant Sud-Ouest d'une colline (calcaire de l'Eocène), à une altitude d'environ 150 m, dans la commune de Saint Gély du Fesc, au lieu-dit Puech du Mas du Juge.

La végétation initiale était constituée par une formation ligneuse basse de *Quercus coccifera*, ayant 1 à 1,50 m de hauteur et un recouvrement de 90 à 95%. Dans la classification de l'école Braun-Blanquet, cette végétation est à rattacher à l'association *Cocciferetum* Br.–Bl. 1924, sous-association *brachypodietosum* Br.–Bl. 1935 (Braun-Blanquet et al. 1952).

Le choix de la station a été motivé par l'existence de nombreuses études antérieures (Long et al. 1958, 1961, Trabaud 1962, Poissonet 1966). Ces études permettaient d'avoir une bonne connaissance des lieux dès le début des expériences, donc de délimiter une zone relativement homogène, d'une surface suffisante. De plus, la proximité de Montpellier facilitait les tâches matérielles.

Le choix des traitements

Pour essayer de mieux cerner le rôle de l'action de l'hom-

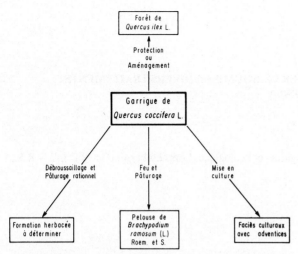

Fig. 1. Schéma des évolutions d'une végétation de garrigue, raisonné en hypothèse de travail, pour les expériences du Puech du Mas du Juge.

me sur cette garrigue, les expériences envisagées avaient pour but de simuler séparément chacune des actions prépondérantes que sont le feu, le pâturage et la mise en culture. L'étude de chacune de ces actions a été comparée à des végétations-témoins, protégées de toute intervention humaine et de l'action des herbivores tels que moutons, lapins... De plus, quelques facteurs du milieu ont été étudiés; ce sont l'érosion et le ruissellement en relation avec les données météorologiques locales, l'activité biologique des sols, leur bilan hydrique...

Les hypothèses de travail formulées sont schématisées sur la figure 1 et peuvent être résumées ainsi: la garrigue de *Quercus coccifera*, provenant de la forêt primitive de *Quercus ilex*, est susceptible par l'action conjuguée du feu, de l'abandon de la culture et du pâturage:

– de retourner à la forêt, par protection ou mise en défens, ou par aménagement sylvicole;

– d'évoluer vers une pelouse de *Brachypodium ramosum* par l'action répétée du feu et du pâturage;

– d'évoluer vers une formation herbacée dont la composition est à déterminer, par l'action du pâturage rationnel (Long et al. 1964);

– d'évoluer vers un stade à adventices par la mise en culture.

Deux de ces expériences sont présentées ci-après: celle avec l'action répétée du feu et celle avec des coupes successives simulant le pâturage.

L'action répétée du feu

Afin de mettre en évidence l'effet des conditions saison-

nières, les mises à feu sont réalisées à deux époques de l'année: au printemps: P (mai) avant le départ de la végétation du *Quercus coccifera* et en automne: A (septembre) à la fin de sa période de végétation.

La fréquence des feux sur six années est la suivante:

– un feu tous les deux ans, donc 3 feux en six ans (3),

– un feu tous les trois ans, donc 2 feux en six ans (2),

– un feu tous les six ans (1).

Cette dernière fréquence paraît correspondre à une périodicité courante dans la région. Les fréquences plus rapprochées tendent à accélérer l'évolution.

L'expérience 'feu' est une expérience factorielle; en conséquence, les combinaisons des deux facteurs contrôlés ('époque de mise à feu' et 'fréquence des feux') correspondent à six traitements. Ces traitements sont répétés 5 fois.

Les coupes successives

L'expérience 'coupe' a été faite pour simuler un effet 'pâturage':

– par débroussaillage mécanique, en 1969, la végétation broyée ayant été laissée sur place,

– par des coupes ultérieures à un rythme de pâture. En fait, l'absence totale d'animaux entraîne l'absence de piétinement, de restitutions, ce qui est évidemment critiquable; mais elle est rendue nécessaire, dans une première phase, pour des raisons de dimensions des parcelles expérimentales.

L'étude de l'action 'coupe' inclut deux facteurs contrôlés:

– un facteur 'date de coupe', afin de suivre, dans la mesure du possible, le rythme de croissance de la végétation,

– un facteur 'fertilisation' pour apprécier les potentialités des espèces herbacées présentes.

Les dates de coupe sont fixées de la manière suivante:

– la coupe J a lieu à une hauteur moyenne de végétation de 15 cm,

– la coupe J + 7, sept jours après la coupe J,

– la coupe J + 14, quatorze jours après la coupe J.

Les niveaux de fertilisation sont les suivants:

– absence de fertilisation (F1),

– 100 unités (kg) de N, 100 de P_2O_5 et 100 de K_2O par hectare et par an (F2),

– 200 unités (kg) de N, 200 de P_2O_5 et 200 de K_2O par hectare et par an (F3).

L'expérience est une expérience factorielle; en conséquence, les combinaisons des deux facteurs contrôlés 'dates de coupe' et 'fertilisation' correspondent à neuf traitements. Ces traitements sont répétés 7 fois.

Le dispositif expérimental

Le dispositif comprend des parcelles de 50 m² (10 mX5m), réparties dans un ensemble de deux hectares. La végétation est étudiée périodiquement, le long d'une ligne de 10 m, située à égale distance des deux bords. Ci-après, ne sont présentés que les résultats de l'inventaire des espèces recensées sous la ligne, au printemps, pendant les cinq premières années des expériences, de 1969 à 1974. D'autres données (mesures pondérales, listes d'espèces,...) sont recueillies sur ce dispositif, mais elles ne sont pas étudiées ici.

Les résultats ci-après concernent:
– l'action 'feu' dans quatre blocs de sept parcelles comprenant chacun une parcelle-témoin et les six traitements 'feu', soit 28 parcelles,
– l'action 'coupe' dans deux blocs de neuf parcelles comprenant chacun les neuf traitements 'coupe', soit 18 parcelles.

Les tendances évolutives des cinq premières années

Les tendances évolutives des cinq premières années, extraites de la liste des espèces présentes par ligne, concernent le nombre d'espèces par ligne et la fréquence de chaque espèce dans les 46 lignes étudiées.

Le nombre d'espèces

Le nombre d'espèces varie au cours des années, selon les traitements. Les résultats sont exprimés en désignant par:
– N: le nombre d'espèces recensées chaque année par ligne de 10 m, $\bar{N}$ la valeur moyenne des N par traitement ou par facteur;
– R: le rapport, exprimé en %, entre le nombre d'espèces d'une ligne l'année d'observation et le nombre d'espèces de cette ligne en 1969, avant le début de l'expérience, $\bar{R}$ la valeur moyenne des R par traitement ou par facteur. Le coefficient de sécurité de ces résultats est de 95%.

Les parcelles-témoins (tableau I)

En 1969, les parcelles-témoins, comme toutes les parcelles du dispositif, ont un nombre d'espèces très variable selon les lignes de 10 m. En 1974, le nombre d'espèces tend à être moins variable entre parcelles-témoins qu'en 1969. Ainsi, les parcelles-témoins qui avaient le moins d'espèces par

Tableau 1. Nombre d'espèces dans les parcelles-témoins, par ligne de 10 m, en 1969 et en 1974.

date \ Numéro des parcelles-témoins	77	85	101	106	Valeur moyenne
1969 (N	19	11	14	19	15,7
(R	100	100	100	100	100
1974 (N	20	17	16	18	17,7
(R	105	155	114	95	117,2

N = nombre d'espèces

$$R = \frac{N}{N_{1969}} \times 100$$

ligne à l'origine, présentent le plus grand accroissement de ce nombre.

En fait, le nombre d'espèces par ligne varie selon les années dans chacune des parcelles-témoins, il est soit en augmentation, soit constant, soit en diminution. Cependant, la tendance entre deux années consécutives et dans une parcelle n'est pas obligatoirement celle des autres parcelles. C'est pourquoi il n'est pas exclu une variation cyclique pluri-annuelle du nombre d'espèces, indépendante des conditions climatiques; mais six années d'observation ne sont pas suffisantes pour préciser cette hypothèse.

Les facteurs contrôlés

Après mise à feu ou après débroussaillage, la diminution immédiate du nombre d'espèces par ligne est évidente. Le nombre d'espèces qui restent (s'il en reste) juste après l'application de ces facteurs et son augmentation ultérieure dépendent de la richesse floristique antérieure. C'est pourquoi, comme il a été déjà indiqué, les résultats sont exprimés non seulement en valeur absolue mais encore en valeur relative par rapport à 1969 avant traitement. Pour l'instant, il n'est pas possible de corriger ces chiffres par la variation pluri-annuelle notée dans les parcelles-témoins.

Les facteurs contrôlés dans l'expérience 'feu' (tableau II et figure 2)

Quelle que soit l'époque de mise à feu, l'année suivante, le nombre moyen d'espèces par ligne est souvent voisin ou plus faible que celui des autres années, surtout au printemps. Cependant, en 1974, cinq ans après le début de l'expérience, la valeur de $\bar{N}$ par traitement tend à être supérieure à celle observée en 1969; les espèces apparues sont

Tableau 2. Évolution du nombre moyen d'espèces observées, au printemps par ligne de 10 m, selon les traitements "feu".

Traitements \ Années	1969		1970	1971		1972		1973		1974
1P	14,2	△	13,6	14,5		15,2		16,5		17,7
2P	12,5	△	11,0	10,0		11,7	△	10,7		13,0
3P	11,0	△	12,9	12,8	△	12,8		13,8	△	11,7
1A	15,3	△	11,5	15,0		17,2		16,0		17,3
2A	12,8	△	13,2	14,5		16,2	△	11,7		17,0
3A	13,1	△	13,3	14,2	△	15,3		15,4	△	16,2

a) moyenne ($\bar{N}$) des valeurs absolues (N)

Traitements \ Années	1969		1970	1971		1972		1973		1974
1P	100	△	96	102		107		116		125
2P	100	△	88	80		94	△	86		104
3P	100	△	117	116	△	116		125	△	106
1A	100	△	75	98		112		105		113
2A	100	△	103	113		127	△	91		133
3A	100	△	102	108	△	117		118	△	124

b) moyenne ($\bar{R}$) des valeurs relatives ($\dfrac{N}{N_{1969}} \times 100$)

1P végétation brûlée une seule fois au printemps (1969)
2P végétation brûlée deux fois au printemps (1969, 1972)
3P végétation brûlée trois fois au printemps (1969, 1971, 1973)
1A végétation brûlée une seule fois en automne (1969)
2A végétation brûlée deux fois en automne (1969, 1972)
3A végétation brûlée trois fois en automne (1969, 1971, 1973)
△ années de mises à feu, selon les traitements.

essentiellement des espèces étrangères au *Cocciferetum*. Cette augmentation est la plus forte avec les traitements 1P, 2A et 3A, elle est la plus faible avec les traitements 2P et 3P.

Pour les fréquences de mise à feu, les parcelles ne devant être brûlées qu'une seule fois (1P et 1A) présentent, en moyenne et quelle que soit l'année, un nombre d'espèces par ligne plus élevé que celui des parcelles devant être brûlées plusieurs fois (2P, 2A, 3P, 3A). Néanmoins, en valeur relative, les résultats ne diffèrent pas significativement entre eux bien que, généralement, les résultats des parcelles brûlées tous les deux ans (3P, 3A) tendent à être supérieurs aux autres.

En examinant l'époque de mise à feu, nous observons, en moyenne, plus d'espèces dans les parcelles soumises aux feux d'automne que dans les parcelles soumises aux feux de printemps, en 1971, 1972 et 1974. Par contre, en valeur relative, cette différence est tamponnée par le fait que, à l'origine, les parcelles brûlées à l'automne avaient en moyenne plus d'espèces que les parcelles devant être brûlées au printemps. Cet accroissement du nombre d'espèces après les feux d'automne peut être dû au fait que les feux de printemps (mai) sont allumés au moment où *Quercus coccifera* est en période de croissance et qu'il donne, tout de suite après le feu, des repousses qui couvriront assez rapidement l'espace nu, laissant ainsi peu de place

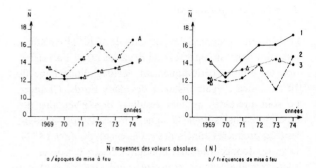

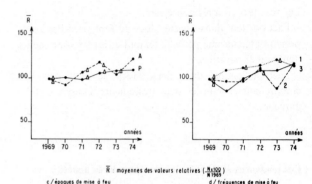

Fig. 2. Évolution du nombre moyen d'espèces observées au printemps, par ligne de 10 m selon les facteurs contrôlés 'feux'. △ mise à feu.

pour l'installation ou le développement d'autres espèces dont le cycle de végétation est virtuellement terminé. Au contraire, en automne (septembre), *Q. coccifera* est en état de vie ralenti ; son aptitude à donner des rejets est moindre, il occupera donc moins d'espace après les feux, ce qui permettra à d'autres espèces de s'installer ou de se développer de l'automne au début du printemps suivant, avant le départ en végétation du *Q. coccifera*.

Les facteurs contrôlés dans l'expérience 'coupe' (tableau III et figure 3)

Immédiatement après le débroussaillage, le nombre moyen d'espèces par ligne correspond à celui observé dans la première strate (0–5 cm) avant débroussaillage, soit environ 60 % des espèces notées dans toutes les strates ; ce sont des espèces enracinées, pour la plupart, sous la ligne d'observation ou aux environs immédiats. L'année suivante (1970), le nombre moyen d'espèces par ligne augmente dans huit traitements sur neuf ; cette augmentation correspond à l'extension des espèces herbacées due aux coupes répétées. En 1971, il n'y a pas de tendance nette

Tableau 3. Evolution du nombre moyen d'espèces observées au printemps, par ligne de 10 m, selon les traitements "coupe"

Traitements	Années	1969 avant débrous-saillage	1969 1ère strate	1970	1971	1972	1973	1974
F1, J		16,0	11,5	10,5	15,0	15,0	15,5	18,5
F1, J+7		14,0	8,5	13,5	11,0	14,5	16,0	16,5
F1, J+14		13,5	6,0	11,0	14,5	17,0	19,5	21,5
F2, J		15,5	9,5	13,5	12,0	17,0	15,5	18,5
F2, J+7		12,0	8,0	12,0	14,5	17,5	20,0	18,0
F2, J+14		12,0	7,5	12,0	11,0	16,0	21,0	21,0
F3, J		16,0	8,0	13,5	7,5	13,0	15,0	16,0
F3, J+7		15,5	10,5	14,0	13,0	16,5	19,0	17,5
F3, J+14		17,5	9,5	14,0	14,5	18,0	19,0	20,5

a) moyenne ($\bar{N}$) des valeurs absolues (N)

Traitements	Années	1969 avant débrous-saillage	1969 1ère strate	1970	1971	1972	1973	1974
F1, J		100	72	66	94	94	97	116
F1, J+7		100	60	99	77	105	114	118
F1, J+14		100	47	83	111	133	149	161
F2, J		100	62	91	83	114	102	125
F2, J+7		100	67	100	121	146	167	150
F2, J+14		100	63	100	92	133	175	175
F3, J		100	49	85	47	82	93	100
F3, J+7		100	68	90	85	108	124	115
F3, J+14		100	53	81	85	103	110	119

b) moyenne ($\bar{R}$) des valeurs relatives ($\dfrac{N}{N_{1969}} \times 100$)

F1: absence de fertilisation
F2: 100 unités de N, 100 de P205 et 100 de K20/ha/an
F3: 200 unités de N, 200 de P205 et 200 de K20/ha/an
J: première date de coupe
J+7: deuxième date de coupe (7 jours après J)
J+14: troisième date de coupe (14 jours après J).

par rapport à 1970. Cependant, le nombre moyen d'espèces par traitement est supérieur à celui de 1969 après débroussaillage dans huit traitements sur neuf. En 1972 et 1973, sept traitements ont plus d'espèces qu'en 1969 avant débroussaillage. Les deux autres traitements sont F1, J et F3, J. En 1974, tous les traitements ont, au moins, autant d'espèces par ligne qu'en 1969 avant débroussaillage.

Les trois dernières années (1972, 1973, 1974) sont marquées par l'apparition de composées des genres *Lactuca, Sonchus, Crepis, Picris, Senecio*... favorisées par la coupe et la fertilisation éventuelle.

Les moyennes $\bar{R}$ des valeurs relatives montrent des tendances comparables à celles des moyennes des valeurs absolues et permettent de noter, entre autres, que trois traitements présentent des pourcentages d'augmentation égaux ou supérieurs à 50%.

Les résultats obtenus par niveau de chaque facteur ont une allure générale sigmoïdale de 1969 après débroussaillage jusqu'à 1974. Des résultats comparables ont été obtenus dans d'autres expériences *in situ* (Poissonet 1974).

Le nombre moyen ($\bar{N}$) d'espèces n'est pas différent entre les trois niveaux du facteur 'fertilisation', pour une même année. Par contre, la moyenne $\bar{R}$ des valeurs relatives $\dfrac{N}{N_{1969}}$ montre une nette différence entre les trois niveaux F1, F2, F3; chaque année, de 1971 à 1974, la valeur moyenne la plus forte est obtenue à partir des résultats des parcelles ayant reçu la fertilisation F2 et la plus faible à partir de ceux des F3. Il en résulte que la relation entre $\bar{R}$ et la fertilisation est parabolique, pour une même année. De telles relations paraboliques sont connues dans certaines expérimentations sur cultures (Lecompt 1965).

En 1974, l'augmentation du nombre d'espèces est de :
– 30% dans les parcelles non fertilisées (F1),
– 50% dans les parcelles moyennement fertilisées (F2),
– 10% dans les parcelles fortement fertilisées (F3).

La fertilisation permet à un certain nombre d'espèces de s'implanter le long de la ligne d'observation (différence entre F1 et F2). De plus, elle joue un rôle sur le développement des individus des espèces réagissant bien à l'engrais, chaque individu occupant plus de place; le nombre des espèces est alors d'autant moins important par ligne que la fertilisation est plus élevée; la richesse floristique le long

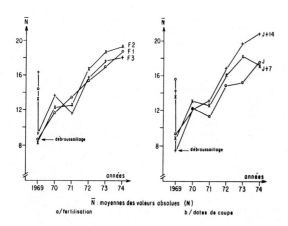

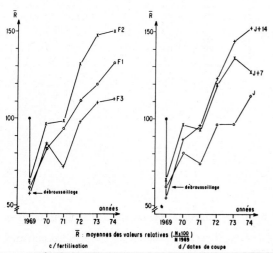

Fig. 3. Évolution du nombre moyen d'espèces observées au printemps, par ligne de 10 m selon les facteurs contrôlés 'coupe'.

de la ligne d'observation en est affectée (différence entre F2 et F3).

Entre les trois dates de coupe J, J + 7, J + 14, le nombre d'espèces varie en 1973 et 1974. Les différences sont beaucoup plus nettes lorsqu'elles sont exprimées en moyennes $\bar{R}$ des valeurs relatives; pour une même année, les valeurs obtenues avec J sont plus faibles que celles obtenues avec J + 7, elles-mêmes étant plus faibles que celles obtenues avec J + 14. Il en résulte que, pour une même année, la relation entre $\bar{R}$ et la date de coupe est curviligne. Cette augmentation du nombre d'espèces semble liée au nombre de jours qui s'écoulent entre les dates de coupe; elle s'accentue avec les années. Les observations présentées sont faites au printemps, saison où la végétation est la plus active; les coupes les plus tardives permettent d'inventorier plus d'espèces tout en laissant à

ces dernières plus de chances de se maintenir et de se multiplier.

Discussion

De 1969 à 1974, le nombre moyen d'espèces par parcelle est, le plus souvent, en augmentation. Cela peut être expliqué par l'ouverture de la végétation au moyen du feu et de la coupe qui permettent le développement d'espèces quasi-absentes dans une garrigue dense de *Quercus coccifera*. Les témoins eux-mêmes, dont les dimensions sont relativement faibles (5 × 10 m), peuvent être influencés par le dispositif expérimental qui les entoure.

De plus, les parcelles-témoins présentent d'importantes fluctuations inter-annuelles. Il en résulte que, sur les six années, leur nombre moyen d'espèces $\bar{N}$ et leur moyenne des rapports $\bar{R}$ ne sont pas différents de ceux des expériences 'feu' et 'coupe'.

En 1974, les moyennes ($\bar{N}$ et $\bar{R}$) des parcelles-témoins ne peuvent être considérées comme différentes de celles des expériences 'feu' et 'coupe', sauf de celles des niveaux F2 et J + 14 qui ont des valeurs $\bar{R}$ supérieures à celles des parcelles-témoins. Par contre, les valeurs $\bar{N}$ et $\bar{R}$ sont plus élevées dans l'expérience 'coupe' que dans l'expérience 'feu'.

Observations sur le comportement de quelques espèces

La permanence de certaines espèces

La plupart des espèces inventoriées à l'origine persistent au cours des cinq années d'observation, quel que soit le traitement appliqué. C'est le cas des espèces constituant la majeure partie de la biomasse initiale telles que *Quercus coccifera, Brachypodium ramosum, Dorycnium pentaphyllum* ssp. *pentaphyllum* ou d'espèces communes dans la garrigue telles que *Rubia peregrina* et *Teucrium chamaedrys*.

L'accroissement de la fréquence de certaines espèces

Des espèces présentes à l'origine deviennent plus fréquentes sous les lignes, quels que soient les facteurs contrôlés; ce sont: *Bromus erectus, Carex halleriana, Centaurea pectinata, Cephalaria leucantha, Galium pumilum, Sedum sediforme*. D'autres espèces ont une fréquence qui augmente le long des lignes dans les parcelles soumises aux feux alors qu'elle diminue dans les parcelles coupées; ce sont: *Asphodelus cerasifer, Bupleurum rigidum, Rubus ulmifolius, Sanguisorba minor. Brachypodium pinnatum* ssp. *phoeni-*

Tableau 4. Espèces nouvelles observées le long de lignes de 10 m, après l'action du "feu" et de la "coupe".

ESPECES COMMUNES AUX ACTIONS "FEU" ET "COUPE"*

Espèces apparaissant définitivement après feux		Espèces apparaissant de façon fugace après un feu		Espèces apparaissant après deux ou trois feux	
Hippocrepis comosa	71	Allium sphaerocephalum	70	Sonchus oleraceus	70
Silene italica	72	Seseli montanum	70	Senecio vulgaris	71
		Avena bromoïdes	71	Althaea hirsuta	73
		Ononis minutissima	71	Daucus carota	74
		Viola alba	71		
		Crepis sancta	72		
		Narcissus juncifolius	74		
		Plantago lanceolata	74		
		Tragopogon porrifolius	74		

ESPECES N'APPARAISSANT QU'AVEC L'ACTION "COUPE"

1970	1971	1973	1974
Dactylis glomerata	Hieracium pilosella	Bromus madritensis	Eryngium campestre
		Convolvulus cantabrica	Geranium rotundifolium
		Crepis pulchra	Odontites lutea
		Geranium robertianum	Osyris alba
		Gladiolus segetum	Reichardia picroïdes
		Lactuca serriola	Taraxacum officinale
		Picris echioïdes	
		Picris hieracioïdes	
		Teucrium montanum	

*L'année d'apparition dans les traitements "coupe" est indiqué par les 2 derniers chiffres, après le nom d'espèces.

coïdes tend à être plus fréquent après un feu, puis sa fréquence diminue si les feux sont répétés. Dans les parcelles coupées, il a tendance à se développer au cours du temps.

La diminution de fréquence de certaines espèces

Cistus monspeliensis et Euphorbia characias ont tendance à être moins fréquent au cours des années, le long des lignes. Il en est de même de Genista scorpius sauf dans les parcelles brûlées au printemps. Hieracium murorum, Phyllyrea angustifolia, Smilax aspera sont moins fréquemment rencontrés le long des lignes dans les parcelles traitées, aussitôt après application des traitements. Néanmoins, dans les parcelles brûlées, leur fréquence augmente par la suite pour atteindre, en 1974, la même fréquence qu'au début de l'expérience.

La disparition de certaines espèces

Quercus ilex disparaît dès la première coupe et après plusieurs feux. Juniperus oxycedrus disparaît dès le premier traitement.

L'apparition de certaines espèces

De 1970 à 1974, des espèces apparaissent le long des lignes dans les deux expériences 'coupe' et 'feu'. Ces espèces figurent au tableau IV. La moitié de ces espèces sont communes aux deux expériences; les autres n'apparaissent qu'avec la 'coupe' et d'abord dans les parcelles fertilisées.

Conclusions

Après cinq années d'expérience, l'action répétée du feu n'a pas permis de faire évoluer la garrigue de Quercus coccifera étudiée vers une une pelouse à Brachypodium ramosum, ce qui n'est pas conforme à l'hypothèse formulée avant la mise en place de l'expérience. Par contre, les coupes successives, avec ou sans fertilisation, ont conduit à une formation à herbacées dominantes dans laquelle persiste la plupart des ligneux du Cocciferetum, conformément à l'hypothèse formulée.

La stabilité de la flore d'origine est remarquable malgré l'augmentation du nombre d'espèces par ligne, la présence d'espèces nouvelles et la modification de la fréquence de certaines espèces par les traitements. L'apparition d'espèces nouvelles le long des lignes provient, en grande partie, d'espèces rares ou peu fréquentes dans la formation étudiée, avant traitement, ou d'espèces des formations environnantes.

Summary

In 1969, several experiments were carried out in a *Quercus coccifera* garrigue, in order to clear up the role of man's action. Two of these experiments are described here:
- one based on the repeated action of fire with two controlled factors (period of fire setting and fire frequency).
- the other simulating rational grazing after mechanical scrub-clearing, with two controlled factors (fertilization and cutting period).

The experimental results presented here are those obtained from 1969 to 1974, by observation along lines. They concern the species number and the behaviour of certain species; it appears from these results:
- Whatever the experiment and the treatment, the number of species increases from 1969 to 1974.
- The number of species is higher with 'autumn fire' than with 'spring fire', and with 'fire every six years' than with other fire frequencies (two and three years).
- The number of species is the highest with mean fertilization and the latest cutting time.
- The number of species in a reference line varies very much year after year, it follows that only the increase of the number of species with mean fertilization and the latest cutting time is higher than that of the reference ones.

The stability of the original flora is noteworthy but the species frequency is modified. A few species appear, these ones are rare in the *Quercus coccifera* garrigue or come from the surrounding vegetation.

The experiment 'fire' has not yet allowed to obtain by succession, the *Brachypodium ramosum* sward -hypothesis currently supported – the experiment 'cutting' has led towards a formation in which the grasses predominate. These experiments are now in progress.

Références bibliographiques

Braun-Blanquet, J., N. Roussine & R. Nègre. 1952. Les groupements végétaux de la France méditerranéenne. Centre National de la Recherche Scientifique, Paris, 297 pp.

Fournier, P. 1961. Les quatres flores de France. Lechevalier, Paris, 1105 pp.

Godron, M.,P. Daget, L. Emberger, G. Long, E. Le Floc'h, J. Poissonet, C. Sauvage & J.-P. Wacquant. 1968. Relevé méthodique de la végétation et du milieu. Code et transcription sur cartes perforées. Centre National de la Recherche Scientifique, Paris 292 pp.

Lecompt, M. 1965. L'expérimentation et les engrais. S.P.I.E.A., Paris, 91 p.

Long, G., J. Rami & L. Visona. 1958, 1961. La végétation du domaine de Coulondres (Hérault). Relation avec les problèmes de mise en valeur. Série des études locales, C.N.R.S., C.G.V., Montpellier, 48 p. (1958) et Bull. Inst. Bot. Univ. Catania, 3(1), 5–52 (1961).

Long, G., F. Fay. & M. Thiault. 1964. Possibilité d'utilisation de la garrigue par le mouton. Journées C.E.T.A., mouton, étude n° 982, 6 pp.

Poissonet, P. 1966. Etude méthodologique en écologie végétale à partir de photographies aériennes. Thèse de spécialité (Biologie Végétale, Ecologie), Faculté des Sciences de Montpellier, 106 pp.

Poissonet, P. 1974. Etude expérimentale sur les 'effets de voisinage' dans des communautés végétales herbacées mono- à pluri-spécifiques. Thèse de Doctorat ès Sciences Naturelles, Université des Sciences et Techniques du Languedoc, Montpellier, 237 pp.

Trabaud, L. 1962. Monographie phytosociologique et écologique de la région de Grabels-Saint Gély du Fesc. Thèse de 3ème cycle, Ecologie, Faculté des Sciences de Montpellier, 131 pp.

Tutin, T. G., V. H. Heywood, N. A. Burges, D. M. Moore, D. H. Valentine, S. M. Walters, D. A. Webb. 1964, 1968, 1972, 1976. Flora Europaea, 4 vol. Cambridge University Press, 1:XXXII, 370 p. (1964), 2:XXVII, 455 p. (1968), 3:XXIX, 370 p. (1972), 4:XXIX, 505 p. (1976).

DAUERQUADRAT-UNTERSUCHUNGEN IN PFLANZENGESELLSCHAFTEN MIT ALPINEN UND MONTANEN RELIKTPFLANZEN IM TRANSDANUBISCHEN MITTELGEBIRGE (UNGARN)*,**

I. ISÉPY

Egyetemi Botanikus Kert, Hort. Bot. Univ., Illés u. 25, H-1083 Budapest, Hungary

Einleitung

Seit etwa 15 Jahren setzt sich in der Geobotanik anstelle der deskriptiven Beschreibung einzelner Pflanzengesellschaften die vielseitige Erforschung von Ökosystemen immer mehr durch. Sichtbarster Ausdruck hierfür sind die IBP- und MAB-Programme der UNESCO, bei denen die Untersuchung der Stoffproduktion und des Stoffumsatzes im Vordergrund stehen. Beispiele für diese komplexe Ökosystemforschung sind in Ungarn die Bearbeitung der Sodaböden des Alföld, der Steppenrasen und Eichenwälder im Donau-Theiss-Gebiet, der *Quercus cerris*-*Wälder* des Ungarischen Mittelgebirges und der Biotope des Balatonsees. Diese vorwiegend ökologisch ausgerichtete Forschung schließt auch die Beobachtung der Sukzession auf Dauerflächen mit ein, da ohne die Kenntnis der Vegetationsdynamik kein umfassender Überblick über die Funktion der Ökosysteme möglich ist. Daneben wurden aber auch dort Dauerbeobachtungsflächen eingerichtet, wo seltene oder gar vom Aussterben bedrohte Arten vorkommen. Diese Standorte sind nicht nur für die Wissenschaft, sondern auch für den Naturschutz von Interesse. Als Beispiel für diese Sukzessionsstudien soll hier über Dauerflächen-Untersuchungen berichtet werden, die an den Dolomithängen des Transdanubischen Mittelgebirges durchgeführt werden, wo subalpine und montane Reliktpflanzen vorkommen.

Etwa 60 Arten, d.h. nur 3,2% der ungarischen Flora sind als alpine, alpin-balkanische oder karpatische Florenelemente anzusehen. Viele davon konnten sich als Glazialrelikte nur infolge des extremen Mikroklimas und der besonderen Geomorphologie der Dolomithänge gegenüber den Mitbewerben bis zum heutigen Tag behaupten (Gams

* Nomenklatur nach Sóo (1964–1972).
** Contribution to the Symposium of the Working Group for Succession Research on Permanent Plots, held at Yerseke, the Netherlands, October 1975.

1936, Zólyomi 1942). Zu nennen sind hier das inselartige Vorkommen der karpatisch-endemischen Art *Primula auricula* subsp. *hungarica*, der dealpin-borealen *Bupleurum longifolium*, *Calamagrostis varia*, *Carduus glaucus*, *Festuca amethystina* und *Moehringia muscosa* sowie der montanen ostbalkanisch-dazischen *Anthyllis calcicola*, *Coronilla vaginalis* und *Galium austriacum* subsp. *balatonicum*, die sowohl die geschlossenen Dolomitfelsrasen (*Festuco-Brometum*) als auch die Karstmischwälder (*Fago-Ornetum*), besonders aber deren Übergangszone besiedeln (Boros 1954, Isépy 1970, Zólyomi 1942, 1960).

Untersuchungsmethode

Von *Primula auricula* subsp. *hungarica* sind in Ungarn nur fünf Fundorte bekannt, von denen das Vorkommen im Fáni-Tal des Vértes-Gebirges für die Anlage von Dauerflächen ausgewählt wurde. Im Übergang vom geschlossenen *Festuco-Brometum* zum *Fago-Ornetum* wurden hier zwei Transekte von 2 m Breite und 25 m Länge in 50 × 50 cm große Quadrate unterteilt und über fünf Jahre für jedes dieser Klein-quadrate eine vollständige Liste der Gefäßpflanzen erstellt sowie der Deckungsgrad der Krautschicht geschätzt. Das gleiche geschah auf jeweils einer Kernfläche im Felsrasen (Größe 5 × 5 m) und im Walde (Größe 10 × 10 m). Diapositivaufnahmen sowie die Kartierung einzelner Quadrate ergänzten die Untersuchungen.

Untersuchungsergebnisse

Die Vegetationsverhältnisse der *Primula auricula* subsp. *hungarica*- Standorte im Fáni-Tal des Vértes-Gebirges sind in Fig. 1 schematisch dargestellt. Danach erstreckt sich zwischen dem bereits seit langem beschriebenen *Festuco-Brometum* (A) und *Fago-Ornetum* (B) (Sóo 1970, Zólyomi

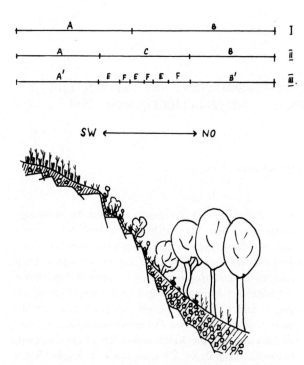

Fig. 1: Vegetationsprofil des *Primula auricula* subsp. *hungarica*-Standortes im Vértes-Gebirge. Gliederung der Vegetationseinheiten nach der Größe der Aufnahmeflächen (I: 100 m², II: 25 m², III: 0.25 m²).
A = *Festuco-Brometum*, B = *Fago-Ornetum*, C = Übergangszone, A′ = geschlossene Felsrasen mit *Bromus erectus* und *Carex humilis*, E = offene Felsrasen mit *Festuca pallens* und *Primula auricula* subsp. *hungarica*, F = Saum mit Sträuchern, B′ = Karstmischwald mit *Fagus sylvatica*, *Fraxinus ornus* und *Carex alba*.

1950) eine Übergangszone C, die sich auf grund der Kleinquadrataufnahmen noch weiter untergliedern ließ (Isépy 1970). Neben den Kümmerformen von Bäumen und Sträuchern der *Fago-Ornetums* (F) finden sich gerade in der Überganszone jene offenen Rasen, die *Primula auricula* subsp. *hungarica* bevorzugt (E).

Auf grund der extremen Boden- und Reliefverhältnisse war es nicht weiter überraschend, daß innerhalb der fünf Beobachtungsjahre keine wesentlichen Vegetationsverschiebungen eintraten. Es ist vielmehr anzunehmen, daß die steilen Felsnasen mit ihrer starken Bodenerosion heute dieselbe Vegetation aufweisen, wie sie in der Kiefern-Birken-Zeit vor etwa 10.000 Jahren auch für viele Standorte in ebener Lage typisch war. Dort, wo in Felsbändern bei geringerer Hangneigung eine minimale Bodenbildung möglich war, entstanden die geschlossenen Rasen des *Festuco-Brometums*. Auf den kolluvialen Talböden ent-

wickelten sich später Buchenwälder. Heute gehen sie hangaufwärts mit abnehmender Bodentiefe in Karstmischwälder über, in denen sich bei höherem Wärmegenuß als im Buchenwald typische wärmeliebende und trockenheitsertragende Arten der Eichenmischwaldzeit (5–6.000 v. Chr.) wie *Digitalis grandiflora*, *Fraxinus ornus*, *Valeriana collina*, *Amelanchier*- und *Cotoneaster*-Arten bis heute behaupten konnten.

Neben dem Mikroklima und dem Boden spielt auch die Lichtverteilung eine wichtige Rolle bei der kleinräumigen Verbreitung der Reliktpflanzen. *Primula auricula* subsp. *hungarica* ist eine Sonnenpflanze, die bei zunehmender Beschattung durch Bäume verschwindet. Entsprechendes gilt auch für den Farn *Asplenium fontanum*, der nach am Ende des vorigen Jahrhunderts im Fáni-Tal häufig war, aber seit 1932 verschwunden ist, nach Jávorka (1940) eine Folge der fortschreitenden Bewaldung des Dolomithanges. *Carduus glaucus* überlebt im Waldsaum, während die Schattenpflanze *Moehringia muscosa* überall im Karstwald zu finden ist. Die montanen, jedoch nicht ausgesprochen subalpinen Arten *Coronilla vaginalis*, *Phyteuma orbiculare* und *Polygala amara* subsp. *brachyptera* sind wiederum in den Felsrasen regelmäßig vertreten.

Um die Konkurrenzverhältnisse zu analysieren, wurden schließlich Korrelationsberechnungen zum Vorkommen der Reliktarten und der dominanten bzw. häufigen Arten in den Kleinquadraten durchgeführt. Dabei schälte sich deutlich das gemeinsame Vorkommen der Reliktelemente in der Übergangszone zwischen dem *Festuco-Brometum* und dem *Fago-Ornetum* heraus. Hier vermindert sich infolge veränderter Standortsfaktoren die Konkurrenzkraft der Wald- und Rasenarten so stark, daß dann die weniger wettbewerbsfähigen subalpinen montanen Arten verstärkt auftreten können. Die interspezifische Konkurrenz der Reliktarten untereinander ist der Übergangszone wichtiger als der Wettbewerb mit den weitverbreiteten Rasen- und Waldpflanzen, da hier nur Arten mit etwa gleichen ökologischen Anspüchen aufeinandertreffen.

Zusammenfassung

Fünfjährige Dauerbeobachtungen von Transekten und Einzelquadraten an Dolomithängen des Transdanubischen Mittelgebirges (Ungarn) ergaben, daß neben den bereits beschriebenen Rasen- (*Festuco-Brometum*) und Waldgesellschaften (*Fago-Ornetum*) eine Übergangszone auftritt, in der montane und subalpine Reliktpflanzen vermehrt vorkommen. Extreme Boden-, Mikroklima- und Lichtverhältnisse sorgten in dieser Zone dafür, daß sich hier diese konkurrenzschwachen Arten bis heute behaupten konnten.

Summary

Five years of investigations on permanent transects and single quadrats in the Transdanubian Mountains (Hungary) showed that a transition zone with many mountain and subalpine relict species occurs between the well-known grassland (*Festuco-Brometum*) and forest (*Fago-Ornetum*). Owing to the extreme soil, microclimate and light conditions of the studied dolomite slopes only here these species of low competitive value could survive.

Literatur

Boros, Á. 1954. A Vértes, a Velencei-hegység, a Velencei-tó és környékük növényföldrajza. Földr. Ért. 3: 280–300.

Gams, H. 1936. Über Reliktföhrenwälder und das Dolomitphänomen. Berlin.

Isépy, I. 1970. Zönologische Untersuchungen im Ost-Vértes-Gebirge. Acta Bot. Hung. 16: 59–110.

Jávorka, S. 1940. Die Entdeckung der Art Asplenium fontanum (L.) Bernh. in Ungarn. Mat. és Term. tud. Ert. 59: 998–1003.

Soó, R. 1964–1972. A magyar flóra és vegetáció rendszertani-növényföldrajzi kézikönyve. I-V. – Synopsis systematico-geobotanica florae vegetationisque Hungariae. I-V Akadémiai Kiadó. Budapest.

Zólyomi, B. 1942. Die Mitteldonau-Florenscheide und das Dolomit-phänomen. Bot. Közlem. 39: 209–231.

Zólyomi, B. 1950. Les phytocénoses des montagnes de Buda et le reboisement des endroits dénudés. Acta Biol. Hung. 1: 7–67.

Zólyomi, B. 1960. Neue Klassifikation der Felsen-Vegetation im pannonischen Raum und der angrenzenden Gebiete. Bot. Közlem. 53: 49–54.

ÄNDERUNGEN IN DER STICKSTOFFVERSORGUNG AUF DAUERFLÄCHEN IM BRACHLAND*·**

Wolfgang SCHMIDT

Lehrstuhl für Geobotanik der Universität Göttingen, Untere Karspüle, D-3400 Göttingen, B.R.D.

Keywords:
Nitrogen mineralization, Old-field vegetation, Permanent plots, Soil nutrients, Vegetation dynamics.

Einleitung

Seit 1968 wird die Vegetationsentwicklung auf einer ursprünglich vegetationsfreien Ackberfläche im Neuen Botanischen Garten der Universität Göttingen experimentell verfolgt (Schmidt 1975). Die Versuchsfläche (0,37 ha) ist in 5 Blöcke unterteilt (Fig. 1), die sich durch ihre Vorbehandlung im Frühjahr 1968 unterscheiden. Bei Block I wurde der gesamte Boden bis in 30 cm Tiefe durch einen dieselgetriebenen Flammenofen gegeben und etwa eine Stunde hitzesterilisiert. Block II wurde mit einem Herbizid (Trapex) vorbehandelt, Block III,

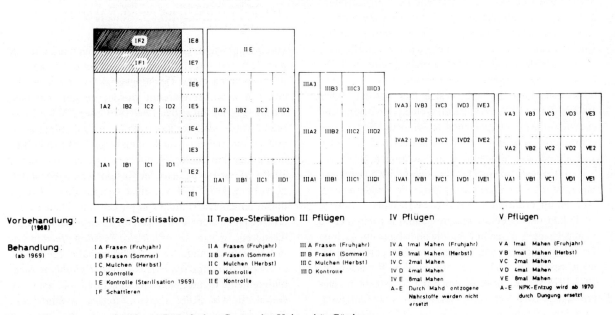

Fig. 1. Versuchsanlage im Neuen Botanischen Garten der Universität Göttingen.

*Nomenklatur der Arten nach Ehrendorfer (1973).
**Contribution to the Symposium of the Working Group for Succession Research on Permanent Plots, held at Yerseke, the Netherlands, October 1975.

47

IV und V wurden im Frühsommer 1968 lediglich tief umgepflügt. Die seit Herbst 1968 beginnende Vegetationsentwicklung zeigte, daß bei vorangegangener Hitze-Sterilisation alle im Boden vorhandenen Pflanzenteile am besten abgetötet worden waren, die Herbizid-Behandlung wirkte dagegen weniger total, während bei den gepflügten Flächen zwangsläufig die rascheste Wiederbesiedlung einsetzte. Dies ist einer der Gründe, warum im weiteren nur auf die Vegetationsentwicklung und Stickstoff-Versorgung im Block I eingegangen werden soll.

Ab 1969 wurden die Blöcke in Streifen von 5 m Breite unterteilt und seitdem jeweils folgender, sich jährlich wiederholender Behandlung unterzogen: IA wird im Frühjahr (Ende April) gefräst, d.h. etwa 20 cm tief fein gepflügt; IB wird im Sommer (Ende Juni/Anfang Juli) gefräst; IC wird im Herbst (September) gemulcht, d.h. das oberirdische Pflanzenmaterial wird abgemäht und verbleibt als Streu auf dem Streifen, und ID blieb seit 1968 unberührte Kontrollfläche.

Vegetationsentwicklung

Seit 1969 wurden auf jeder Parzelle (je 2 pro Streifen) 2–3 Vegetationsaufnahmen pro Vegetationsperiode erstellt, indem der Deckungsgrad in Prozent geschätzt wurde. In den Tab. 1–4 ist jeweils der Jahresmittelwert in der Schätzungsskala nach Londo (Schmidt 1974, Londo 1975) angegeben. Die Darstellung der Vegetationsentwicklung beschränkt sich hier auf die Jahre 1969/1970 und 1973/1974, die zum einen der Beginn und ein vorläufiges End (Zwischen-)stadium der Sukzession charakterisieren, zum anderen aus diesen Zeiträumen auch detaillierte Stickstoff-Untersuchungen vorliegen.

Tab. 1 gibt als Beispiel für eine sekundäre progressive Sukzession (Tüxen & Preising 1942) die Wiederbesiedlung der unberührten Kontrollfläche ID wieder. Bereits im 1. Jahr nach Versuchsbeginn (1969) wurden hier 36 Phanerogamen-Arten notiert, die allerdings nur geringe Deckungsgrade erreichten. Der Deckungsgrad stieg danach rasch an und erreichte in der Krautschicht 1973/1974 Werte um 85%. Auch die Artenzahlen zeigten 6 Jahre nach Versuchsbeginn noch keine Sättigung. Eine Strauchschicht wurde erstmals 1973 notiert, während die Kryptogamen (Moose) nach einem Deckungsgrad-Maximum im Jahre 1970 rasch wieder an Bedeutung verloren haben.

Der Artengruppenwechsel ist in Tab. 1–4 nur sehr

Tabelle 1: Entwicklung der Vegetation auf der Kontrollfläche ID. Angegeben ist der mittlere Deckungsgrad pro Vegetationsperiode und Versuchsstreifen (2 Parzellen) in der Schätzungsskala nach LONDO.

Jahr	1969	1970	1973	1974
Deckung der Strauchschicht (%)	-	-	1.7	6.2
Deckung der Krautschicht (%)	4.8	42.5	83.0	86.7
Deckung der Moosschicht (%)	7.5	35.0	8.0	1.3
Artenzahl	36	40	56	60
Conyza canadensis	.1	2		
Stellaria media	.1	.2		
Senecio vulgaris	.1	.2		
Capsella bursa-pastoris	.1	.2		
Sonchus asper	.1	.2	.1	
Arenaria serpyllifolia	.1	.2		.1
Viola arvensis	.1	.2	.1	.1
Epilobium tetragonum	.1	.2	.1	.1
Poa annua	.1	.2	.1	.1
Lactuca serriola		.2		
Taraxacum officinale	.1	.2	2	2
Picris hieracioides	.1	.1	1.2	2
Solidago canadensis	.1	.4	1.2	1.2
Tussilago farfara	.1	.1	1.2	1.2
Poa trivialis			1.2	1.2
Epilobium angustifolium	.1	.4	.7	.4
Epilobium adenocaulon	.1	.2	.7	.2
Epilobium parviflorum	.1	.2	.2	.1
Salix caprea	.1	.1	.2	.2
Betula pendula	.1	.1	.1	.2
Fraxinus excelsior		.1	.1	.2

Außerdem mit einem mittleren Deckungsgrad von weniger als 1 % (.1):

1969: Polygonum persicaria, Anagallis arvensis, Aethusa cynapium, Polygonum aviculare, Sherardia arvensis, Sambucus nigra.
1969/1970: Fallopia convolvulus, Chaenarrhinum minus, Atriplex patula, Matricaria discoidea, Papaver rhoeas, Sonchus oleraceus, Chenopodium album.
1969/1970/1973/1974: Cirsium arvense, Galium aparine, Myosotis arvensis, Plantago major.
1970: Veronica polita.
1970/1973: Epilobium hirsutum.
1970/1973/1974: Equisetum arvense, Tripleurospermum inodorum, Cirsium vulgare, Crepis biennis, Geum urbanum, Acer platanoides.
1973: Bellis perennis, Potentilla spec.
1973/1974: Cerastium holosteoides, Epilobium montanum, Poa palustris, Crepis capillaris, Clematis vitalba, Fragaria vesca, Phleum pratense, Daucus carota, Dactylis glomerata, Cornus sanguinea, Trifolium dubium, Arrhenatherum elatius, Solidago gigantea, Festuca ovina, Senecio jac obea, Poa pratensis, Artemisia vulgaris, Rosa cf. canina, Agrostis stolonifera, Aster cf. novae-angliae, Sorbus aucuparia, Scrophularia nodosa, Luzula luzuloides, Prunella vulgaris, Hordeum murinum, Inula conyza, Deschampsia caespitosa, Rubus fruticosus.
1974: Acer pseudo-platanus, Acer campestre, Crataegus monogyna, Hieracium sylvaticum, Erigeron acris, Lolium perenne, Campanula spec.

vereinfacht wiedergegeben: mit einer eigenen Zeile sind nur die Arten aufgeführt, die im Verlauf der Vegetationsentwicklung einen Deckungsgrad von mehr als 1% erreichten, wobei getrennt wurde, ob der maximale Deckungsgrad entweder 1969/1970 oder 1973/1974 erreicht wurde. Auf der Kontrollfläche ID dominierten in den ersten beiden Jahren einjährige Ackerunkräuter und einige kurzlebige Hemikryptophyten, von denen 1970 *Conyza canadensis* im Gesamtaspekt besonders hervortrat. 1973/1974 war diese Gruppe dann abgelöst durch langlebige Hemikryptophyten und Phanerophyten, die einen mehrschichtigen Pflanzenbestand charakterisieren: *Taraxacum officinale, Tussilago farfara* und *Poa trivialis* dominierten in Erdbodennähe; *Picris hieracioides, Solidago canadensis* und *Epilobium*-Arten bildeten eine dichte, obere Krautschicht, die dann von Gehölzen wie *Salix caprea, Betula pendula* und *Fraxinus excelsior* überragt wurde.

Im Gegensatz dazu fehlte auf der im Frühjahr gefrästen Fläche IA eine Artengruppe, deren maximaler Deckungsgrad bereits zu Beginn des Versuchs lag (Tab. 2). Annuelle und sich vegetativ rasch regenerierende Aus-

Tabelle 2: Entwicklung der Vegetation auf der Fläche IA (Fräsen im Frühjahr).
Angegeben ist der mittlere Deckungsgrad pro Vegetationsperiode
und Versuchsstreifen (2 Parzellen) in der Schätzungsskala nach
LONDO.

Jahr	1969	1970	1973	1974
Deckung der Krautschicht (%)	3.0	11.3	75.0	70.0
Deckung der Moosschicht (%)	0.3	4.8	-	-
Artenzahl	29	39	49	55
Tussilago farfara	.1	.1	3	3
Poa trivialis	.1	.1	.4	.4
Taraxacum officinale	.1	.1	.4	.4
Chenopodium album	.1	.1	.4	.2
Viola arvensis	.1	.1	.2	.4
Conyza canadensis	.1	.2	.2	.2
Poa annua	.1	.1	.2	.2
Cirsium arvense	.1	.1	.2	.2
Sonchus asper	.1	.1	.2	.1
Capsella bursa-pastoris	.1	.1	.1	.2
Galium aparine		.1	.7	.7
Atriplex patula		.1	.7	.2
Sonchus arvensis			.2	.2
Bromus arvensis			.2	.2
Ranunculus repens			.1	.2

Außerdem mit einem mittleren Deckungsgrad von weniger als 1 % (.1):

1969/1970: Chaenarrhinum minus, Salix caprea, Senecio vernalis.
1969/1970/1973: Epilobium adenocaulon.
1969/1970/1974: Clematis vitalba.
1969/1970/1973/1974: Equisetum arvense, Senecio vulgaris, Sonchus oleraceus,
    Fallopia convolvulus, Epilobium angustifolium, Epilobium tetragonum,
    Arenaria serpyllifolia, Stachys palustris, Veronica polita, Stellaria
    media, Myosotis arvensis, Tripleurospermum inodorum.
1969/1973/1974: Plantago major.
1969/1974: Anagallis arvensis.
1970: Salix spec., Betula pendula.
1970/1973: Epilobium parviflorum, Cirsium vulgare.
1970/1973/1974: Lactuca serriola, Polygonum persicaria, Urtica dioica,
    Papaver rhoeas, Picris hieracioides.
1970/1974: Matricaria discoidea.
1973: Dactylis glomerata, Fumaria officinalis, Daucus carota, Trifolium
    pratense, Crepis capillaris.
1973/1974: Solidago canadensis, Agrostis stolonifera, Hordeum murinum,
    Sinapis arvensis, Melilotus alba, Crepis biennis, Cornus sanguinea,
    Rumex crispus.
1974: Veronica hederifolia, Cerastium holosteoides, Crepis pulchra,
    Aethusa cynapium, Silene noctiflora, Euphorbia helioscopia, Thlaspi
    arvense, Festuca pratensis, Festuca ovina, Crataegus cf. monogyna,
    Lolium perenne.

Tabelle 3: Entwicklung der Vegetation auf der Fläche IB (Fräsen im Sommer).
Angegeben ist der mittlere Deckungsgrad pro Vegetationsperiode und
Versuchsstreifen (2 Parzellen) in der Schätzungsskala nach LONDO.

Jahr	1969	1970	1973	1974
Deckung der Krautschicht (%)	5.8	25.0	60.0	68.3
Deckung der Moosschicht (%)	1.3	15.0	-	-
Artenzahl	33	29	42	44
Epilobium parviflorum	.1	.2	.1	
Sonchus oleraceus	.4		.1	.1
Conyza canadensis	.1	1.2	.2	.1
Veronica polita	.1	.2	.1	.1
Epilobium tetragonum	.1	.2	.1	.1
Poa annua	.1	.2	.1	.1
Poa trivialis	.1	.1	2	1.2
Cirsium arvense	.1	.1	.7	.7
Taraxacum officinale	.1	.1	.7	.7
Tussilago farfara	.1	.1	.7	.7
Capsella bursa-pastoris	.1	.1	.4	.1
Senecio vulgaris	.1	.2	.2	.2
Arenaria serpyllifolia	.1	.2	.2	.2
Viola arvensis	.1	.2	.2	.2
Papaver rhoeas	.1	.2	.2	.2
Epilobium angustifolium	.1	.2	.2	.1
Sonchus asper	.1	.2	.1	.2
Equisetum arvense	.1	.1	.2	.1
Myosotis arvensis	.1	.1	.2	.1
Crepis pulchra			.4	1.2

Außerdem mit einem mittleren Deckungsgrad von weniger als 1 % (.1):

1969: Anagallis arvensis, Chaenarrhinum minus, Plantago major, Polygonum
    aviculare, Luzula luzuloides.
1969/1970/1973/1974: Galium aparine, Epilobium adenocaulon, Chenopodium
    album, Tripleurospermum inodorum, Stellaria media.
1969/1974: Fraxinus excelsior, Aethusa cynapium, Sherardia arvensis, Fallopia
    convolvulus.
1970/1973: Veronica hederifolia, Senecio vernalis.
1970/1973/1974: Lactuca serriola, Picris hieracioides.
1970/1974: Epilobium hirsutum.
1973: Senecio jaccobea, Stachys palustris, Solidago canadensis, Prunus
    avium, Hypochoeris radicata.
1973/1974: Crepis capillaris, Veronica persica, Ranunculus repens, Bromus
    arvensis, Festuca pratensis, Atriplex patula, Cerastium holosteoides,
    Sinapis arvensis.
1974: Crepis biennis, Polygonum persicaria, Fumaria officinalis, Lolium
    perenne.

dauernde zeigten vielmehr mit zunehmendem Gesamtdeckungsgrad auch zunehmende Einzelwerte, wobei
seit 1973 besonders *Tussilago farfara* mit seinen großen
Blättern die Fläche fast vollständig überdeckte.

Die Vegetationsentwicklung auf der sommergefräs-
ten Fläche IB war ähnlich (Tab. 3). Allerdings konnten
hier wieder zwei Artengruppen unterschieden werden:
Conyza canadensis und eine Reihe von anuellen Acke-
runkräutern traten 1969/1970 stark in Erscheinung,
während nach 5 bzw. 6 Jahren Arten wie *Poa trivialis,
Cirsium arvense, Taraxacum officinale, Tussilago farfara* usw. vorherrschten. Kennzeichnend für die som-
mergefrästen Flächen waren ferner die niedrigen Ge-
samtartenzahlen während der bisherigen Vegetationsentwicklung, die auf den radikalen Eingriff durch das
Fräsen hinweisen, dem nur relativ wenige Arten widerstehen können.

Die Vegetationsentwicklung auf den gemulchten Flächen ähnelte wiederum sehr stark der der Kontrollfläche (Tab. 4). Das gilt nicht nur für die Zusammensetzung der beiden unterschiedenen Artengruppen, sondern auch für den Gesamtaspekt dieser Fläche nach
dem Ablauf von 6 Versuchsjahren. Eine Strauchschicht fehlte allerdings. In den ersten beiden Versuchsjahren waren die Artenzahlen der gemulchten Fläche

Tabelle 4: Entwicklung der Vegetation auf der Fläche IC (Mulchen im Herbst).
Angegeben ist der mittlere Deckungsgrad pro Vegetationsperiode
und Versuchsstreifen (2 Parzellen) in der Schätzungsskala nach LONDO.

Jahr	1969	1970	1973	1974
Deckung der Krautschicht (%)	6.8	37.5	86.7	86.7
Deckung der Moosschicht (%)	7.3	28.8	4.0	0.7
Artenzahl	43	47	49	49
Sonchus oleraceus	.2	.1		
Atriplex patula	.1	.2		
Conyza canadensis	.1	1.2	.1	
Epilobium angustifolium	.1	.4	.2	.2
Poa annua	.1	.4	.1	.1
Taraxacum officinale	.1	.1	2	2
Poa trivialis	.1	.1	2	2
Picris hieracioides	.1	.1	2	2
Solidago canadensis	.1	.4	1.2	.7
Tussilago farfara	.1	.1	.7	.4
Salix caprea	.1	.1	.2	.4
Epilobium adenocaulon	.1	.2	.2	.1
Poa pratensis			.2	.2

Außerdem mit einem mittleren Deckungsgrad von weniger als 1 % (.1):

1969: Sinapis arvensis, Polygonum persicaria, Anagallis arvensis, Aethusa
    cynapium, Silene noctiflora, Avena sativa, Erechtithes hieraciifolia,
    Tragopogon pratensis, Sambucus nigra.
1969/1970: Stellaria media, Senecio vulgaris, Chaenarrhinum minus, Chenopodium
    album, Sonchus asper, Papaver rhoeas, Capsella bursa-pastoris, Arenaria
    serpyllifolia, Polygonum aviculare, Populus tremula.
1969/1970/1973: Epilobium parviflorum.
1969/1970/1973/1974: Viola arvensis, Epilobium tetragonum, Cirsium arvense,
    Betula pendula, Fraxinus excelsior, Tripleurospermum inodorum, Galium
    aparine, Plantago major, Equisetum arvense, Myosotis arvensis.
1969/1970/1974: Fallopia convolvulus.
1970: Matricaria discoidea, Lactuca serriola, Veronica polita, Gnaphalium
    uliginosum, Cirsium palustre, Senecio vernalis.
1970/1973: Hypochoeris radicata.
1970/1973/1974: Cirsium vulgare, Solidago gigantea, Epilobium hirsutum,
    Clematis vitalba, Crepis capillaris, Geum urbanum.
1973: Trifolium dubium, Holcus lanatus, Urtica dioica, Caltha palustris.
1973/1974: Cerastium holosteoides, Epilobium montanum, Dactylis glomerata,
    Senecio jaccobea, Calamagrostis epigejos, Festuca pratensis, Crepis
    biennis, Hieracium pilosella, Phleum pratense, Rosa cf. canina, Sorbus
    aucuparia, Bellis perennis, Acer platanoides, Aster cf. novae-angliae,
    Cornus sanguinea, Artemisia vulgaris.
1974: Daucus carota, Hieracium sylvaticum, Acer pseudo-platanus, Melilotus
    alba, Sonchus arvensis, Ranunculus auricomus.

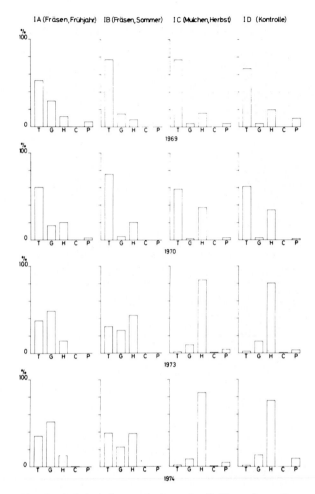

IA (Fräsen, Frühjahr)   IB (Fräsen, Sommer)   IC (Mulchen, Herbst)   ID (Kontrolle)

Fig. Fig. 2. Lebensformen-Spektren für die Versuchsstreifen
kungsgrad. T = Therophyten, G = Geophyten, H = Hemikryptophyten, C = Chamaephyten, P = Phanerophyten. Life
from spectra for the areas IA–ID shown as percentages of total
cover.

am höchsten von allen hier gegenübergestellten Ver-
suchsvarianten, wiesen danach kaum noch eine Steige-
rung auf und blieben 1973/1974 sogar konstant.

Diese kurze Übersicht über die Vegetationsentwicklung soll durch eine Gegenüberstellung der
Lebensformen-Spektren abgeschlossen werden (Fig.
2). Auf den Kontroll- und Mulchflächen stellte sich die
Sukzession in den ersten 6 Jahren weitgehend als ein
Wettbewerb zwischen Therophyten und Hemikryptophyten dar. 1969/1970 betrug der Anteil der Einjährigen
zwischen 60 und 75% am Gesamtdeckungsgrad,
1973/1974 war er auf weniger als 2% abgesunken. Die

Hemikryptophyten herrschten jetzt mit 75–85% eindeu-
tig vor. Neben diesen beiden Lebensformentypen ist die
Rolle der weiteren nur von untergeordneter Bedeutung.
Die Geophyten – es handelte sich ausschließlich um
Rhizomgeophyten – und auch die Phanerophyten erreichten dabei auf den Kontrollflächen etwas höhere
Deckungsgradanteile als auf den gemulchten Flächen,
wo der Anteil der Hemikryptophyten absolut am höchsten war. Auf den gefrästen Flächen besaßen dagegen
die Annuellen 1973/1974 immer noch einen Anteil von
30–40%. Hemikryptophyten hatten sich nur auf der im
Sommer gefrästen Fläche stärker durchgesetzt, während Rhizomgeophyten – insbesondere Tussilago farfara – auf der im Frühjahr gefrästen Fläche IA etwa die
Hälfte des Gesamtdeckungsgrades ausmachten.

Änderungen in der Gesamtstickstoff-Konzentration des Oberbodens

Die Untersuchungsergebnisse zur Änderung der Stick-
stoffversorgung sollen sich auf zwei Fragen konzentrie-
ren:
1. Inwieweit lassen sich im Verlauf der Sukzession Änderungen in der Gesamtstickstoff-Konzentration des
Oberbodens nachweisen?
2. In welchem Maße erfolgten Änderungen in der
Mineralstickstoff-Nachlieferung, d.h. in der Mineralisation von organisch gebundenem Stickstoff zu pflanzen-
aufnehmbarem NH_4^+ und NO_3^-?

Zur Beantwortung der ersten Frage wurden 1969 und
1973 jeweils am Ende der Vegetationsperiode (Oktober)
von jeder Parzelle Bodenproben aus den obersten 0–10
cm des Mineralbodens mit dem Pürckhauer-Bohrer
entnommen (Mischprobe aus 10 Einzelproben) und der
getrocknete und feingemahlene Boden einer Gesamt-
Stickstoff-Bestimmung nach Kjeldahl unterzogen
(Bremner 1960). Die in Tab. 5 dargestellten Untersuchungsergebnisse zeigen für die Streifen IA und IB
(Fräsen im Frühjahr bzw. Sommer) nur minimale Änderungen, für die Flächen IC (Mulchen) und ID (Kontrolle) dagegen eine deutliche Erhöhung in der

Tabelle 5: Gesamtstickstoff-Konzentration in den obersten 0-10 cm des Mineralbodens.
Sicherung der Mittelwertdifferenzen nach dem DUNCAN-Test (n.s. = nicht
signifikant, * P ≤ 10 %, ** P ≤ 5 %, *** P ≤ 1 %).

Behandlung			1969	1973	DUNCAN-Test
IA	Fräsen (Frühjahr)	x̄ (% N)	0.121	0.123	n.s.
		s (% N)	0.007	0.004	
IB	Fräsen (Sommer)	x̄ (% N)	0.127	0.123	n.s.
		s (% N)	0.007	0.000	
IC	Mulchen (Herbst)	x̄ (% N)	0.110	0.124	n.s.
		s (% N)	0.004	0.031	
ID	Kontrolle	x̄ (% N)	0.124	0.138	**
		s (% N)	0.002	0.004	

50

Gesamtstickstoff-Konzentration. Für den Streifen IC lassen sich die Mittelwertdifferenzen allerdings auf Grund der hohen Streuung nicht sichern.

Diese Ergebnisse deuten an, daß bei der Forsetzung der alten ackerbaulichen Behandlung in Form des Fräsens auch das ursprünglich niedrige Gesamtstickstoffniveau des Bodens beibehalten wird. Entwickelt sich der Bestand von der Annuellen- und Geophyten-Gesellschaft weiter zu einem typischen Brachland mit langlebigen Hemikryptophyten und Phanerophyten, so kommt es bereits im Verlauf von 3–4 Jahren zu einer geringfügigen Erhöhung der Gesamtstickstoff-Konzentration im Oberboden. In Böden von Ackerbrachen des Westerwaldes fand Von Borstel (1974) dagegen keinen statistisch sicherbaren Anstieg im Gesamtstickstoffgehalt mit zunehmender Brachedauer, lediglich die Konzentration an organisch gebundenem Kohlenstoff nahm deutlich zu.

Änderungen in der Mineralstickstoff-Nachlieferung

Stickstoff kann von den höheren Pflanzen nur in der Form von NH_4^+ und/oder NO_3^- aufgenommen werden, deren Angebot an ungedüngten Standorten nahezu ausschließlich durch die mikrobielle Zersetzung stickstoffhaltiger organischer Substanz (Mineralisation) bestimmt wird (Ellenberg 1964). Von der gesamten Mineralstickstoff-Produktion (Bruttomineralisation steht der Vegetation nur der Teil zur Verfügung, welcher über den mikrobiellen Bedarf hinaus mineralisiert wird (Nettomineralisation) und nicht durch Auswaschung, Festlegung oder Denitrifikation verloren geht. Um das Stickstoffangebot des Bodens an die Vegetation ökologisch sinnvoll zu erfassen, muß daher die Stickstoff-Nettomineralisation bestimmt werden, indem die Mineralstickstoff-Aufnahme durch die Pflanzen für eine bestimmte Zeit unterbunden wird. Dazu werden in einem sogenannten Brutversuch Bodenmischproben entnommen und ihr aktueller Gehalt an NH_4^+ und NO_3^- bestimmt. Gleichzeitig wird ein Teil der Bodenprobe ohne Veränderung des Wassergehaltes in naturfeuchtem Zustand in Polyäthylenbeuteln verschlossen und an der Entnahmestelle im Boden aufbewahrt. Nach sechswöchiger Bebrütung wird erneut der Mineralstickstoff-Gehalt bestimmt und aus der Differenz zwischen den Anfangs- und Endkonzentrationen an NH_4^+ und NO_3^- kann dann die Nettomineralisation in ppm N pro Woche errechnet werden. Durch die La-

gerung der Brutproben am Entnahmeort werden dessen Temperaturverhältnisse berücksichtigt. Abweichend von den natürlichen Verhältnissen bleibt der Wassergehalt dagegen über den Bebrütungszeitraum von 6 Wochen konstant. Um den Einfluß der Wassergehaltsschwankungen und die jahreszeitliche Dynamik der Mineralisation dennoch zu berücksichtigen, wurden alle 3 Wochen neue Brutproben angesetzt, so daß sich pro Bodenhorizont jeweils zwei Serien überlappen (Runge 1970). Die Analyse des Mineralstickstoffs erfolgte durch fraktionierte Wasserdampfdestillation aus dem Filtrat einer Bodensuspension (Gerlach 1973).

Die Ergebnisse derartiger Mineralisationsuntersuchungen in den Jahren 1970 und 1974 sind für zwei Flächen (IB, ID) in Fig. 3 dargestellt. Danach zeigt die Mineralstickstoff-Nachlieferung 1970 ein mehr oder weniger stark ausgeprägtes Frühjahrsmaximum, was in erster Linie auf einen hohen Anteil an leicht zersetzbarer organischer Substanz und den Anstieg der Bodentemperaturen in dieser Jahreszeit zurückzuführen war. Im Sommer und Frühherbst war die Mineralisation dann häufig mit Schwankungen im Bodenwassergehalt korreliert. Mit sinkenden Temperaturen im Herbst nahm dann auch die Mineralisation wieder ab.

1974 war ein entsprechender Jahresgang nur noch auf den gefrästen Flächen wieder zu beobachten. Die Mulch- und Kontrollflächen zeichneten sich im 6. Versuchsjahr jedoch nicht nur durch eine fehlende Jahresdynamik aus, vielmehr waren auch die Mineralisationsraten absolut sehr viel niedriger als im Jahre 1970, wie die Gegenüberstellung in Tab. 6 zeigt. Ökologisch am sinnvollsten ist es dabei, das Mineralstickstoff-Angebot an die höheren Pflanzen nicht als Summe der Mineralisationsintensität zu vergleichen, sondern diesen Wert mit Hilfe des Bodenvolumengewichts und unter Berücksichtigung der Profiltiefe auf den flächenbezogenen Wert kg N_{min}/ha/30 Wochen umzurechnen. Für 1970 wurde dabei für den Block I ein einheitliches Bodenvolumengewicht von 1,15 g/cm^3 für die obersten 0–20 cm des Mineralbodens ermittelt, während es 1974 je nach Behandlungsmaßnahme größere Unterschiede gab. Danach war das Mineralstickstoff-Angebot 1970 auf den gefrästen Flächen mit 133 (IA) und 139 (IB) kg N_{min}/ha/30 Wochen am höchsten, dann folgte die gemulchte Fläche (IC) mit 116 kg N_{min}/ha/30 Wochen und deutlich niedriger die unbehandelte Fläche (ID) mit 67 kg N_{min}/ha/30 Wochen. 1974 war diese Rangfolge zwar erhalten geblieben, die absoluten Werte zeigten jedoch ein gänzlich anderes Bild. Keine Untersuchungsfläche erreichte die

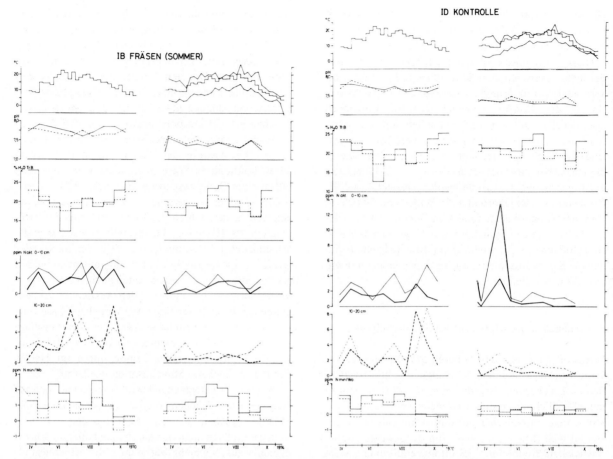

Fig. 3. Jahresgänge der Bodentemperatur (Wochenmittelwerte 10 cm Tiefe 1970/1974, Minimum-/Maximumwerte 1 cm Tiefe nur 1974), des Säueregrads, des Wassergehalts, des aktuellen Mineralstickstoffgehalts und der Stickstoff-Nettomineralisation im Boden der Versuchsflächen IB (Fräsen, Sommer) und ID (Kontrolle). Ausgezogene Linie: 0–10 cm Bodentiefe; gestrichelte Linie: 10–20 cm Bodentiefe; dicke Linie: Nitrat-Konzentration des aktuellen Mineralstickstoffgehalts; dünne Linie: Ammonium-Konzentration des aktuellen Mineralstickstoffgehalts. Fluctuations in soil temperature (weekly means at 10 cm depth 1970/1974, minimum-maximum values at 1 cm depth for 1974 only), pH, water content, mineral nitrogen content and net nitrogen mineralisation in the soils of areas IB and ID, Drawn line: 0–10 cm depth; stippled line: 10–20 cm depth; fat line: nitrate concentration of mineral nitrogen content; thin line: ammonia concentration of mineral nitrogen content.

Tabelle 6: Bodenvolumengewichte (0-20 cm) sowie mittlere Bodentemperatur (10 cm Tiefe, nach Angaben der Wetter-warte Göttingen), mittlerer Bodenwassergehalt (0-20 cm) und Mineralstickstoff-Nachlieferung (0-20 cm) für die Vegetationsperioden 1970 und 1974 (30 Wochen).

Behandlung	Volumengewicht g/cm³		Temperatur °C		Wassergehalt % H₂O		Mineralstickstoff-Nachlieferung			
							ppm N/30 w		kg N/ha/30 w	
	1970	1974	1970	1974	1970	1974	1970	1974	1970	1974
IA Fräsen (Frühjahr)	1.15	1.14	14.5	14.0	20.5	18.3	57.7	42.8	133	98
IB Fräsen (Sommer)	1.15	1.13	14.5	13.7	20.5	19.6	60.3	57.1	139	129
IC Mulchen (Herbst)	1.15	1.31	14.6	13.9	20.9	19.8	50.6	17.2	116	45
ID Kontrolle	1.15	1.23	14.5	13.7	20.1	20.6	29.2	13.3	67	33

Mineralstickstoff-Nachlieferungsrate von 1970, wofür vielleicht die um etwa 0,5°C niedrigere Bodentemperatur im Jahr 1974 verantwortlich sein könnte. Während aber auf der Fläche IB (Fräsen, Sommer) der Mineralisationswert noch nahe beim Wert des Jahres 1970 lag, war er auf den übrigen Flächen stärker abgesunken: am stärksten auf der gemulchten Fläche (IC) und der Kontrollfläche (ID), wo 1974 nur 40–50% des Angebots von 1970 erreicht wurde. Vergleichbare Daten für *Solidago canadensis*-Gesellschaften, die den Pflanzenbeständen der Mulch- und Kontrollflächen in Aufbau und Zusammensetzung sehr ähnlich sind, nennt Kronisch (1975). Er fand dort eine Nachlieferung von 38–86 kg N_{min}/ha/30 Wochen, während die üppigen *Urtica dioica*-Bestände mit meist mehr als 200 kg N_{min}/ha/30 Wochen deutlich besser versorgte Ruderalgesellschaften darstellten.

Diskussion

Die aufgezeigten Unterschiede in der Mineralstickstoff-Nachlieferung und ihre Veränderungen im Laufe der Sukzession können bei einem so komplexen Prozeß wie der mikrobiellen Zersetzung organischer Substanz verschiedenste Ursachen haben. Einige erste Hinweise liefern die Daten in Tab. 6 und die Ergebnisse der vegetationskundlichen Analyse:
1. Auf die 1974 insgesamt niedrigeren Bodentemperaturen und ihren Einfluß auf die Stickstoffmineralisation wurde bereits hingewiesen.
2. Während 1970 der Bodenwassergehalt auf allen vier Versuchsflächen etwa gleich hoch war, zeigte sich 1974 eine deutliche Abstufung. Mit zunehmender Deckungsgrad der Krautschicht (Tab. 1–4) stieg der Bodenwassergehalt im Mittel der Vegetationsperiode von IA nach ID an. Allerdings ergibt sich hier keine direkte Beziehung zur Änderung in der Stickstoffmineralisation.
3. Im direkten Zusammenhang mit den Änderungen in der Stickstoffmineralisation stehen Verschiebungen im Bodenvolumengewicht. Während sich auf den gefrästen Flächen die Bodenvolumengewichte gegenüber 1970 nicht verändert hatten, fand auf den Streifen IC und ID infolge fehlender Bodenauflockerung eine Verdichtung statt. Es ist zu vermuten, daß sich dadurch gleichzeitig das Luftporenvolumen verringerte. Ob dies jedoch allein schon zu einer niedrigeren Mineralisation führte, erscheint zweifelhaft, da Anzeichen von Sauerstoffarmut und Vernässung des Bodens bisher nicht beobachtet werden konnten.

4. Denkbar ist schließlich, daß der Vegetationswechsel selbst die Änderungen in der Stickstoffmineralisation bestimmte. So zeichneten sich ja die gefrästen Flächen auch 1974 noch durch einen hohen Anteil an Therophyten aus (Fig. 2), deren produzierte organische Substanz jährlich vollständig abstirbt und im Boden meist unverzüglich mineralisiert wird. Von den ausdauernden Geophyten und Hemikryptophyten wird ein großer Teil einmal im Jahr durch das Fräsen zerkleinert. Dies geschieht in einer Phase, in der das Pflanzenmaterial durch hohe Stickstoffgehalte ausgezeichnet ist, die dann sehr leicht von den Mikroorganismen verwertet werden können. Ein entsprechender Mineralisationsschub im Frühjahr bzw. Sommer ist in den Jahresgängen der gefrästen Flächen erkennbar (Fig. 3). Anders war die Situation dagegen auf der Mulch- und Kontrollfläche. Hier dominierten Hemikryptophyten, deren oberirdische Pflanzenteile weitgehend absterben oder gemäht werden. Von Vertretern dieser Lebensformengruppe – insbesondere ausdauernden Gräsern und hochwüchsigen Stauden wie *Solidago* –, aber auch Rhizomgeophyten wie *Tussilago* und *Cirsium arvense*, ist bekannt, daß sie am Ende der Vegetationsperiode Stickstoffverbindungen aus ihren absterbenden Organen in überdauernde Pflanzenteile wie Wurzeln, Rhizome, Blattscheiden usw. verlagern (Chwastek 1963, Otzen & Koridon 1970; Hirose 1971, 1974, Wagner 1972). Sie besitzen einen inneren Stickstoff-Kreislauf, der sie von der mikrobiellen Nachlieferung über den Boden weitgehend unabhängig macht. Für tropische Savannen (De Rham 1970), Pfeifengraswiesen (Leon 1968) und Halbtrockenrasen (Wagner 1972) wird ein solcher innerer Stickstoffumsatz vermutet, der meist mit geringen Stickstoffmineralisationsraten im Brutversuch einhergeht. Aus der Umschichtung in den Lebensformen vom Therophyten-Stadium des Jahres 1970 zum Hemikryptophyten-Bestand im Jahre 1974 kann angenommen werden, daß auf den Brachflächen der äußere Stickstoff-Kreislauf der ersten Vegetationsperioden jetzt einem inneren gewichen ist. Ob dies tatsächlich zutrifft, müssen jedoch weitere Untersuchungen (Jahresgänge der Stickstoffmineralisation in späteren Jahren, Stickstoffbilanzen dominierender Arten usw.) zeigen.

Zusammenfassung

Seit 1968 wurde die Vegetationsentwicklung eines ursprünglich vegetationsfreien, hitzesterilisierten Brachlandes auf einem kalkreichen Auelehm verfolgt. Ohne den Einfluß menschlicher Kulturmaßnahmen konnten verschiedene Sukzessionsstadien unterschieden werden, die sich nach floristischen und physiognomischen Gesichtspunkten gliedern ließen. In den ersten sechs Jahren stellte sich die Sukzession weitgehend als ein Wettbewerb zwischen den Therophyten und den Hemikryptophyten dar. Durch verschiedene Kulturmaßnahmen wie jährliches Pflügen im Frühjahr bzw. Sommer oder Mulchen im Herbst konnte die Sukzession so abgewandelt werden, daß Ersatzgesellschaften entstanden.

Fünf Jahre nach Versuchsbeginn hatte der Gesamtstickstoffgehalt im Oberboden der unbehandelten Fläche leicht zugenommen, während die gepflügten Flächen weiterhin auf ihrem niedrigen Niveau verharrten. Ein Vergleich der Stickstoffmineralisation von 1970 mit 1974 zeigte eine deutliche Verminderung der Nachlieferung mit zunehmendem Brachealter (1970: 67 kg N_{min}/ha/30 Wochen; 1974: 33 kg N_{min}/ha/30 Wochen). Demgegenüber hatte den regelmäßig gepflügten Flächen nach fünf Jahren die Mineralstickstoff-Nachlieferung nicht signuifikant abgenommen. Auf Grund dieses Befundes wird vermutet, daß in Ackerbrachen der äußere Nährstoffkreislauf der frühen Therophyten-Stadien später durch einen inneren abgelöst wird, wenn langlebige Hemikryptophyten, Rhizom- und Wurzelknospengeophyten vorherrschen.

Summary

Starting in 1968 plant succession and nitrogen status in the top soil of an old field was investigated. The soil is a calcareous loam which was sterilized by heating. In the absence of human influence distinct successional stages with characteristic floristical and physiognomical features were observed. During the first six years vegetational development was characterized by competition between therophytes and hemicryptophytes. Agricultural treatments such as annual ploughing (spring, summer) and cutting (autumn) were found to change the trend of the succession.

Five years after starting the experiment total nitrogen content in the top soil of the control area had increased slightly, while the ploughed plots persisted in their low values. A comparison of the nitrogen mineralisation between 1970 and 1974 showed decreasing amounts of mineral nitrogen in the later stage (1970: 67 kg N_{min}/ha/30 weeks; 33 kg N_{min}/ha/30 weeks). On the other hand, mineral nitrogen supply on the ploughed plots was not significantly different after the five-years-period. It is suggested that in old-field vegetation the external nitrogen cycle of the early therophyte stage changed to an internal one when long-lived hemicryptophytes, rhizome- and root-budding geophytes became dominant.

Tables 1 to 4 show the vegetation development under different treatments; the figures represent average cover values per vegetation season using Londo's (1975) scale. Table 5 contains the total nitrogen concentration values in the uppermost of 10 cm of mineral soil, and Table 6 lists the values for soil volume, mean soil temperatures, average water content, and mineral nitrogen supplies for 1970 and 1974.

Literatur

Borstel, U.O.v. 1974. Untersuchungen zur Vegetationsentwicklung auf ökologisch verschiedenen Grünland- und akkerbrachen hessischer Mittelgebirge (Westerwald, Rhön, Vogelsberg). Diss. Univ. Gießen, 159 pp.

Bremmner, J.M. 1960. Determination of nitrogen in soil by the Kjeldahl-method. J. Agr. Sci. 55: 11–33.

Chwastek, M. 1963. The influence of nutritional soil resources, especially phosphorus content, on the dominance of Molinia coerulea (L.) Moench in the meadow sward. Poznan Soc. of Friends of Sci. Dept. of Agric. and Sylvic. Sci. 14: 277–356.

Ehrendorfer, F. 1973. Liste der Gefässpflanzen Mitteleuropas. Fischer, Stuttgart.

Ellenberg, H. 1964. Stickstoff als Standortsfaktor. Ber. Dtsch. Bot. Ges. 77: 82–92.

Gerlach, A. 1973. Methodische Untersuchungen zur Bestimmung der Stickstoffnettomineralisation. Scripta Geobotanica 5: 115 pp.

Hirose, T. 1971. Nitrogen turnover and dry matter production of a Solidago altissima population. Jap. J. Ecol. 21: 18–32.

Hirose, T. 1974. On the phosphorus budget of a perennial herb Solidago altissima population. Bot. Mag. Tokyo 87: 89–98.

Kronisch, R. 1975. Zur Stickstoff-Versorgung von Ruderalpflanzen-Gesellschaften in Göttingen. Staatsexamensarbeit Univ. Göttingen, 75 pp.

Léon, R. 1968. Balance d'eau et d'azote dans les prairies à des alentours de Zurich. Veröff. Geobot. Inst. ETH Zürich, Stiftg. Rübel 41: 2–68.

Londo, G. 1975. Dezimalskala für die vegetationskundliche Aufnahme von Dauerquadraten. In: Schmidt, W. (Red.), Sukzessionsforschung. Ber. Int. Sym. IVfV, Rinteln 1973: 613–618. Cramer, Vaduz.

Otzen, D. & A.H. Koridon. 1970. Seasonal fluctuations of organic food reserves in underground parts of Cirsium arvense (L.) Scop. and Tussilago farfara L. Acta Bot. Neerl. 19: 495–502.

de Rham, P. 1970. L'azote dans quelques forêts, savanes et terrains de culture d'Afrique tropicale humide (Côte d'Ivoire). Veröff. Geob. Inst. ETH Zürich, Stiftg. Rübel 45: 1–124.

Runge, M. 1970. Untersuchungen zur Bestimmung der Mineralstickstoff-Nachlieferung am Standort. Flora 159: 233–257.

Schmidt, W. 1974. Bericht über die Arbeitsgruppe für Sukzessionsforschung auf Dauerflächen der Internationalen Vereinigung für Vegetationskunde. Vegetatio 29: 69–73.

Schmidt, W. 1975. Vegetationsentwicklung auf Brachland – Ergebnisse eines fünfjährigen Sukzessions-Versuches. In: Schmidt, W. (Red.), Sukzessionsforschung. Ber. Int. Symp. IVfV, Rinteln 1973: 407–434. Cramer, Vaduz.

Tüxen, R. & E. Preising. 1942. Grundbegriffe und Methoden zum Studium der Wasser- und Sumpfpflanzen-Gesellschaften. Deutsche Wasserwirtschaft 37: 3–22.

Wagner, P. 1972. Untersuchungen über Biomasse und Stickstoffhaushalt eines Halbtrockenrasens. Dipl.Arb. Univ. Göttingen, 54 pp.

EXPERIMENTAL SUCCESSION RESEARCH IN A COASTAL DUNE GRASSLAND, A PRELIMINARY REPORT*, **

E. VAN DER MAAREL***

Division of Geobotany, University of Nijmegen, Toernooiveld, 6525 ED Nijmegen, The Netherlands.

Keywords:
Dune grassland, Grazing, Mowing, Permanent plot, Rabbit, Succession, Trampling, Voorne dunes

Introduction

The research introduced here forms part of a long-term succession study in a dune grassland complex of ca. 2000 sq. m near the Biological Station 'Weevers' Duin', Oostvoorne, in the mediaeval zone of the Voorne dunes (see van der Maarel & Westhoff 1964, van der Maarel 1966a, b, 1975a for a general description of the area). The dune grassland was used for grazing until the 1920's and afterwards as a golflinks. Since 1940 the area is hardly used. Until 1950 very extensive grazing occurred, whilst since the start of the Biological Station in 1952 the area is regularly trodden by investigators. Furthermore a small population of rabbits, burrowing in neighbouring dunes has been grazing the grassland.

In 1963 this grassland complex was described in detail with multivariate methods on the basis of over 1000 relevés resulting in a vegetation map 1 : 200 (van der Maarel 1966b). In addition a plan was developed for a combined study of both the spatial and temporal variation of the complex. This plan involved the yearly recording of floristic composition and structure in some permanent transsects on the basis of the 1963 vegetation map.

In 1967 a first report on the study, dealing with the relation between pattern and process was taken up in the contribution by van der Maarel & van Leeuwen (1967) to the Symposium of the International Society for Vegetation Science on Syndynamics at Rinteln. Unfortunately, the

proceedings of this symposium are still not available. Therefore, some of the results from this contribution are presented here. In 1966 the analysis was reorganized because of the changes in boundaries in the vegetation mosaic as compared with the 1963 situation. From that year onwards ca. 40 permanent plots of 2×2 m have been followed annually. This change in approach was elucidated in a second contribution on the project (van der Maarel 1975b). Since the 1963 analysis was not restricted to relevés of homogeneous spots but also included species lists of each sq. m of a grid laid down in the are, it was possible to compare 4 sq. m data from later years with the 1963 situation.

In 1970 the grid was analyzed on a 4 sq. m basis for 50 % of the total area, which enabled a close comparison with the overall situation of 1963. The main change in the area was an increase in height and cover of taller grasses (van der Maarel 1975b). It was therefore decided to start with some management experiments, in order to find out whether this development is a gradual reaction on the termination of the regular management. Such a development might be impeded by increased pressure on the grassland, or promoted by a further decrease in pressure. The experiments include yearly mowing, exclusion of trampling and exclusion of both trampling and rabbit grazing. Some general observations on the experimental and control quadrats (each measuring again 4 sq. m) will be presented here in addition to the information given in the 1967 study on the relation between pattern and process. A comprehensive account on the study, concentrating on the relation between spatial and temporal variation in floristic structure, will be prepared in a few years, after the direction of development in the experimental plots will have become clear.

* Nomenclature of vascular plants follows Heukels-van Oostroom (1975), that of syntaxa Westhoff & den Held (1969).
** Contribution to the Symposium of the Working Group for Succession Research on Permanent plots, held at Yerseke, The Netherlands, October 1975.
*** Dedicated to Prof. Dr. Heinrich Walter on the occasion of his 80th birthday.

Phytosociological characteristics of the 1963 situation

The dune grassland which is studied in detail consists of a mosaic of vegetation types which gradually merge into each other and also form a continuum from a syntaxonomical point of view (see van der Maarel 1966b for a detailed description). Figs. 1–2, which will be further discussed below show the general situation.

There are two small dune tops, one in the SE part and one in the NW corner (see height contours). The highest spots were covered with an open *Corynephorus canescens* community belonging to the *Erodio-Koelerion* in which winter annuals like *Erodium glutinosum* and *Phleum arenarium* occurred. On the slopes a low closed grassland vegetation was found with *Festuca tenuifolia*, *Agrostis tenuis* and *Anthoxanthum odoratum* as the prominent grasses, *Galium verum*, *Lotus corniculatus*, *Thymus pulegioides*, *Potentilla tabernaemontani*, *Euphrasia officinalis* and *Hieracium pilosella* amongst the many herbs, and *Hypnum cupressiforme* Hedw. var. *lacunosum* Brid and *Rhytidiadelphus squarrosus* (Hedw.) Warnst, as the dominant mosses. This community complex can be assigned to the *Festuco-Galietum maritimi*.

Further down the W and NW slope of the larger dune in the SE part a taller grassland of the same community complex occurred with *Helictotrichon pubescens* and *Achillea millefolium*.

The depression in the W. part was characterized by a dense vegetation of *Calamagrostis epigejos* with *Molinietalia*-species like *Lysimachia vulgaris* and *Lythrum salicaria*. In wet winters and springs the phreatic water table may rise to a few dm below the surface here.

The flat part of intermediate height is covered by an open and low vegetation with *Plantago coronopus*, *Sagina nodosa*, *Cerastium semidecandrum* and *Erophila verna*, thus syntaxonomically forming a transition between *Nanocyperion flavescentis* and *Erodio-Koelerion*. This part has a partially artificial, i.e. shell-containing substrate which is trampled comparatively strongly. A path running through this area is bordered by a *Lolio-Plantaginion* vegetation.

In the total area of investigation of 2000 sq. m no less than 126 species of vascular plants were found in which the families *Poaceae*, *Fabaceae* and *Asteraceae*, are represented with 20, 10 and 10 % respectively. The most interesting (i.e. regionally or nationally rare) species are *Gentianella campestris* ssp. *baltica*, *Centaurium minus*, *Eleocharis quinqueflora*, and *Sagina apetala* ssp. *erecta*, all locally characteristic of the transition between the *Festuco-Galietum* and *Nanocyperion* communities.

In order to cope with the two fold difficulty of delineating syntaxonomical units and of discerning discontinuities in the field, a method of phytosociological characterization by means of syntaxonomical species groups was developed (see Westhoff & van der Maarel 1973 for the method as such and van der Maarel 1975a, b for the application in this succession project). The species were, as far as possible, assigned to a syntaxonomical unit, usually on the alliance level. For each 4 sq. m quadrat the quantitative contribution of all species for each syntaxonomical unit was calculated as % of the total sum of species contributions. Fig. 1 a gives the results for 1963. Squares with the same composition as to prominent species groups are taken together, delineated towards neighbouring groups and drawn in according to the height contour pattern. The resulting map may be called a vegetation cartogram, i.e. a synthesized picture which includes only the most important elements of a total diagram of the vegetation composition (Westhoff 1965). The map units are recognizable from the combination of letters symbolizing the groups. Capitals denote groups with high absolute representation or a high ranking, lower cases refer to lower representation or ranking (see fig. 1).

Phytosociological changes from 1963-1970

Fig. 1 b gives the syntaxonomical cartogram for the year 1970. The most conspicuous change concerns the increase in the contribution of *Arrhenatherion* and *Mesobromion* species, i.e. species of taller and more closed grasslands, particularly on the slopes. This increase is brought about mainly by *Arrhenatherum elatius*, *Galium mollugo*, *Helictotrichon pubescens* and *Ononis repens*. On the moister sites the *Molinio-Arrhenatheretea* species (notably *Holcus lanatus*, *Poa pratensis* and *Vicia cracca*) increased. The *Nanocyperion* species of moist open sites, e.g. *Sagina nodosa*, decreased. The *Festuco-Sedetalia* and *Erodio-Koelerion* species, decreased on the drier places, e.g. *Phleum arenarium*. Increased on the moist open sites, e.g. *Erophila verna*, where they replaced more or less the *Nanocyperion* contribution. The general interpretation of these changes may be: 1) a slight desiccation of the area and 2) a gradual succession from the heavily managed to less affected grassland types. The desiccation effect may be partially of a temporary kind, since the precipitation in May and June 1970 was only 28.3 mm against 95.1 mm on the average (although the rest of the summer was wetter than average). There is also a more definitive form of desiccation, caused

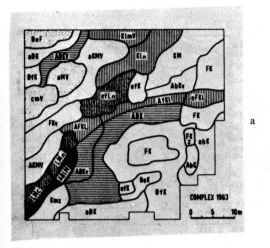
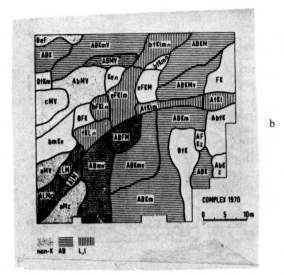

Fig. 1. Vegetation map of the dune grassland succession research area based on analyses of squares of 4 sq.m., with units of equal composition as to prominent syntaxonomical species groups. a. Situation 1963. b. Situation 1970 (van der Maarel 1975a).

A	Arrhenatherion species group contributing	> 20% or ranking	1 or 2
a	Arrhenatherion species group contributing	15–20%	
B	Mesobromion	> 20% ,,	1 or 2
b	Mesobromion	15–20%	
F	Festuco-Sedatalia	> 20%	
f	Festuco-Sedatalia	15–20%	
e	Erodio-Koelerion	1–10%	
K	Galio-Koelerion	> 20% ,,	1 or 2
L	Lolio-Plantaginion	> 10% ,,	2
M	Molinio-Arrhenatheratea	> 20% ,,	1 or 2
m	Molinio-Arrhenatheretea	15–20%	
n	Nanocyperion	1–10%	
V	Violion caninae	> 20% ,,	1 or 2
v	Violion caninae	15–20%	
Z	Berberidion and Trifolio-Geranietea	10–20%	

by the extraction of water towards neighbouring horticultural grounds (van der Maarel & Westhoff 1964).

Interestingly, the increase of shrubs is only very little, although there is plenty of *Hippophae rhamnoides*, *Crataegus monogyna* and *Salix repens* scrub in the surroundings.

From the year-to-year observations from 1963–1966 it became clear that apart from a possible succession, i.e. a directional change, fluctuations in structure and floristic composition occur, which apparently are determined by fluctuations in the amount of precipitation, and connected with this in the yearly course of the groundwater table. The general trend is as follows: In wet years (particularly in years with moist spring and early summer conditions) the vegetation structure on the slopes and in the depression is dense and species of the *Violion caninae* group, e.g. *Anthoxanthum odoratum*, *Briza media* and *Sieglingia decumbens*, occur with high cover values; on the flat central part the *Nanocyperion* species are prominent. The two

summer annuals *Centaurium minus* and *Gentianella campestris* ssp. *baltica* do react even more spectacularly on early summer rain. They germinate in May-June and flower in August-September. In dry years vegetation remains low and open, giving room to *Festuco-Sedetatia* species on the slopes and *Erodio-Koelerion* species on the dune tops and the flat central part, whereas *Violion caninae* and *Nanocyperion* species decrease.

Relation between spatial and temporal patterns of variation

Two approaches have been followed so far. A more complete treatment, in which all aspects of spatio-temporal relations will be treated, will be taken up in the comprehensive account announced in the introduction.

A first approach is to follow the overall pattern of species diversity (van der Maarel 1975a). Species numbers of

59

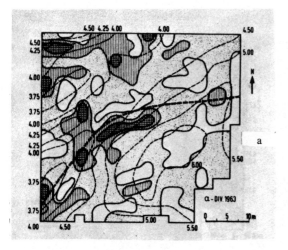
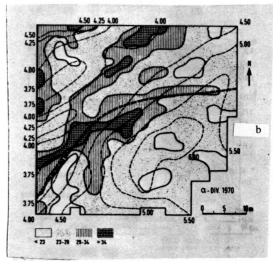

Fig. 2. Diversity pattern in the same area based on species counts in 4 sq.m. squares. a. Situation 1963, b. Situation 1970.

vascular plants, mosses and lichens were determined in all 4 sq. m quadrats available in the 1963 and 1970 analyses. (Fig. 2). The maximum values for 1963 and 1970 were 41 and 42, the minimum values 15 and 19 respectively.

The diversity pattern is drawn in on the basis of isolines for species numbers in four intervals. The 1963 pattern shows zones of high diversity at intermediate heights (height contours, figures are in m above N.A.P. = Dutch Ordnance Level, are taken from van der Maarel 1966b). The path and its borders cause an increase in species richness at heights below 4,50 m. This should be interpreted in relation to the average groundwater table, which is found at ca 3.80 m in winters of wet years and at 3.40 m in wet summers.

The diversity pattern of 1970 is largely the same. Although the circumstances in this year were rather different, particularly concerning the summer drought on one hand and the general development towards the taller grassland syntaxa on the other hand, both maxima and minima and the distribution of species richness are very similar.

In the second approach (van der Maarel & van Leeuwen 1967) year-to-year changes from 1963–1966 in a transect were analyzed (fig. 3). The transect is situated at the western edge of the study area (fig. 1). The height contours presented in fig. 2 can be used as a reference; the transect ends towards the path. The pattern of homogeneous vegetation spots as delineated in 1963 (van der Maarel 1966b) is taken as a basis (fig. 3f). The vegetation types indicated here are local types as distinguished by van der Maarel (1966b). They can be easily interpreted with the help of the syntaxonomical indication presented above.

In addition fig. 3e shows a map with the structural types as distinguished on the basis of height and cover of the various growth forms involved (van der Maarel 1966b, see also van der Maarel 1969 for an ordination of these structural types).

The species diversity pattern (fig. 3a) is based on a diversity measure which was especially devised for this study (van der Maarel, 1966b). It is indicated as α_v and is calculated for each homogeneous spot from $\alpha_v = S \cdot {}^e \log A$, where S is number of species per spot and A its area, expressed in a unit area depending on the structural type. (The unit area is determined as an area on which less than 10 % of the total species number is found). The α_v measure appeared to be highly correlated with the α-index of Williams (1964) based on plant unit counts. For the structural types involved in this area the unit area was 0.1 sq. dm). The diversity pattern is similar to that presented in fig. 2a. In addition we find a relation between diversity and structural type. A simplified chi square test showed significant association, P < 0.05, between diversity values $\alpha_v > 2.7$ and structural types 1–4, i.e. the open to semi-closed types.

Fig. 3b shows the pattern of change. As 'instability measure' we used the average value of $I_T/I_{T\max}$ over the successive years. I_T follows from $S_2 - S_c + S_1 - S_c$, where S_1 and S_2 are numbers of species in successive years and S_c is the number of constant species. $I_{T\max}$ equals $S_1 + S_2$. Note that this measure is equal to the complement to 1 of Sörensen's similarity quotient (cf. van der Maarel 1969 and particularly Londo 1971 for similar measures). The pattern suggest a relation with the diversity and indeed a

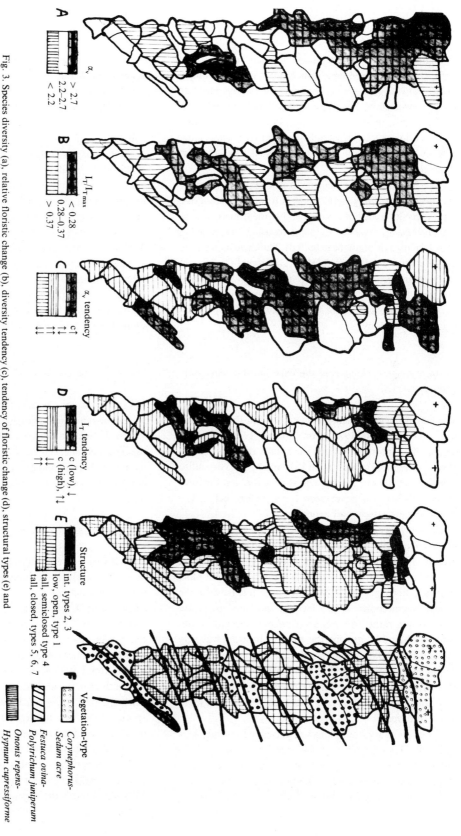

Fig. 3. Species diversity (a), relative floristic change (b), diversity tendency (c), tendency of floristic change (d), structural types (e) and vegetation types (f) in a transect within the dune grassland succession research area (van der Maarel & van Leeuwen 1967). See text for further explanation.

A α_r
> 2.7
2.2–2.7
< 2.2

B $I_T/I_{T\,max}$
< 0.28
0.28–0.37
> 0.37

C α_r tendency
c↑
c↑↓
↓↓

D I_T tendency
c (low), ↓
c (high), ↑↓
↑↑

E Structure
int. types 2, 3
low, open, type 1
tall, semiclosed type 4
tall, closed, types 5, 6, 7

F Vegetation-type
Corynephorus–
Sedum acre
Festuca ovina–
Polytrichum juniperum
Ononis repens–
Hypnum cupressiforme
Festuca ovina–
Rhytidiadelphus
squarrosus
Anthoxanthum odoratum–
Sieglingia decumbens
Calamagrostis epigejos–
Vicia cracca
Lolium perenne–
Cynosurus cristatus

chi square test for association between $\alpha > 2.7$ and $I_T/I_{T\,max} < 0.28$ showed significance at $P < 0.05$. One might argue that with high species numbers the relative measure $I./I_{T\,max}$ will show low figures. Therefore the test was repeated towards I_T and again an association was found, be it at a lower level of significance and with a less clear pattern in zones with variation in structure at short distances.

Figs. 3 c and d show tendencies of change in diversity and instability. The general trend is towards keeping the same diversity level, but at either end of the transect a zone with increasing diversity is found (causing higher levels of instability than might be expected). The trend concerning instability is towards lower levels. At the top of the transect at the transition between the *Corynephorus-Sedum acre* and the *Ononis repens-Hypnum cupressiforme* type we find a zone of increasing change, which coincides with decreasing species diversity.

Phytosociological changes in the experimental plots (preliminary results)

The management experiments mentioned in the Introduction were started in 1970 (mowing) and 1971 (other experiments). Unfortunately for the results the local rabbit population started to incerase in 1971 and by 1975 the entire area was heavily overgrazed, whilst even the first burrows appeared (they had never been found before here). The heavy grazing effected all experiments.

The mowing experiment is carried out in a part of dune grassland complex situated just W of the investigation area described above (fig. 1). One small strip is yearly mown (in winter or early spring) and three pairs of permanent plots were laid down on either side of the borderline between the mown and unmown areas. Two of the three pairs are situated in a rather low *Festuco-Galietum* vegetation. Because of the rabbit grazing the effect of mowing was very limited here. Only in a damp *Calamagrostis epigejos-Lysimachia vulgaris* vegetation the effect of mowing was clear: the removal of the dense *Calamagrostis* mat offered herbs like *Lysimachia vulgaris* and *Lythrum salicaria* more space to develop.

The effect of non-trampling was traceable for some years in a tall *Calamagrostis epigejos* vegetation, but by 1975 differences between trampled and non-trampled sites were largely wiped out because of the intensive rabbit grazing.

The only clear effect obtained was that of preventing rabbits to graze, especially in the flat part with an open and

low vegetation and the surrounding low *Festuco-Galietum* vegetation. Within the experimental plots here we find a significantly denser plant cover and a more varied species composition. This tendency is most pronounced for the plot in the central flat area where both the *Nanocyperion* and the *Festuco-Galietum* species increased.

To our surprise a seedling of *Calluna vulgaris* appeared in this plot already in 1972. The plant has grown up ever since and in 1975 it had reached a height of ca 10 cm and formed some branches (cf. Gimingham 1975 for the development of *Calluna*). The species occurs in a small and declining population in the same grassland complex at a distance of 300 m and at some small sites in another dune grassland complex at a distance of 600 m. Although establishment in the investigation area is not unexpected in view of the soil conditions, no single seedling had been found during the preceding years of intensive research, and no single further seedling have been noticed after 1972.

Discussion and conclusions

The observations described here may be discussed and should be continued with respect to two general themes: 1. The distinction between effects of environmental fluctuations and successional trends in the interpretation of phytosociological changes. 2. The relation between pattern and process in the dynamics of spatial vegetation complexes.

Fluctuation and succession

As to fluctuation and succession the following can be summarized. Clearly the development towards taller grasslands with *Arrhenatherion* and *Mesobromion* species is a successional trend. If put in terms of environmental dynamics (van Leeuwen 1966, van der Maarel 1976) we may interpret this succession as a reaction on a decrease in the environmental dynamics (cf. Sloet van Oldruitenborgh 1976).

This trend is effected, but not seriously bent off by fluctuations in moisture conditions. Those fluctuations are still very well traceable, notably in the lower areas and species of the *Nanocyperion* and *Agropyro-Rumicion scrispi* do clearly respond to them.

The successional trend towards taller grassland is nevertheless susceptible to a more or less natural factor, i.e. the ever growing influence of the local rabbit population since 1971. As a result the vegetation cover is becoming lower and more open. The exclusion experiments, intended to show the influence of moderate rabbit grazing are now developing a refugium character for the types existing until 1972!

Another aspect of the exclusion experiment concerns the establishment of shrubs. The general absence of saplings (seedlings, mainly of *Crataegus monogyna* and *Rhamnus cathartica*, occur but all die off soon) does not seem to be determined by grazing or trampling since the exclusion quadrats do not contain saplings either. There is one case of spread of an already established individual of *Hippophae rhamnoides* in an exclusion quadrat and here an influence may be claimed.

Both for saving the entire grassland complex and for maintaining the proper intention of the exclusion and mowing experiments it would be necessary to reduce the rabbit population. This is not easy, since the study area, although belonging itself to the Biological Station, is freely accessible from nearby horticultural grounds and a large dune nature reserve.

The results of the ordination of the 1963 and 1970 data on the 40 permanent quadrats selected for continuous research (described in van der Maarel 1975a, b) suggest that the phytosociological trends described above are reasonably detected. Still a repetition of the overall quadrat analysis as carried out in 1970 (on the basis of which figs. 1 and 2 were composed) within a few years may be rewarding, certainly if the area would have recovered from the present rabbit stress.

In order to understand the species-to-species reactions on the moisture fluctuations it would be effective to vary the amount of precipitation. Such experiments, which are difficult to realize in more or less natural ecosystems are possible here and planned for 1977. (In the meantime they started early 1978).

Pattern and process

As to the relation between pattern and process it is obvious that analyses of permanent transects, i.e. contiguous or at least transect-wise located permanent plots, provide useful material for spatio-temporal studies of vegetation. The 1970 lay-out of permanent quadrats and their size, 4 sq. m, seem to satisfy the aim of this study, but again studying a more extensive area would be most helpful.

The results presented in fig. 3 have been discussed only briefly. It should be stressed that the relations described between diversity, floristic changes and structure need to be checked by more careful analyses of the entire series of year-to-year data. Van der Maarel & van Leeuwen (1967)

stressed the possible significance of demonstrating the relation between diversity and stability in view of the Relation Theory of van Leeuwen (1966), also van der Maarel 1966a, 1976) and of considerations by MacArthur (1955) and Margelef (1958). This was prior to the torrent of considerations on the subject which has been influencing ecological literature ever since. We do not go into the discussions here, but may still repeat the general interpretation by van der Maarel & van Leeuwen (1967): In landscape types with well-developed environmental gradients and a long history of more-or-less constant management, vegetation zones with high diversity, high stability and intermediate structure will occur in the middle parts of existing gradients. On the contrary, in young and dynamical landscapes strong environmental changes may cause the largest changes in vegetation in the middle parts of existing gradients, whereas the extreme parts, being permanently more dynamic, will be relatively stable in composition. Of course, both general trends may be found together in stable landscapes suddenly subjected to relatively strong changes. The dune grassland under study may well be an example of this complex situation.

In the more comprehensive paper announced in the introduction this interpretation will be elaborated in view of the recent literature on diversity-stability relations.

Summary

Some results are presented on a long-term succession research in a dune grassland complex in which experiments concerning mowing, grazing and trampling are involved. The complex consists of local communities of the *Erodio-Koelerion*, *Festuco-Galietum maritimi*, and *Nanocuperion* with elements of *Molinietalia*, *Arrhenatheretalia*, *Meso-bromion*, *Violion caninae* and *Agropyro-Rumicion crispi*. It is analysed with permanent transects, i.e. series of permanent quadrats forming series along transects, and through detailed mapping of the area at intervals of 7–10 years. Changes in floristic composition are described in terms of both fluctuation and succession and interpreted in view of environmental dynamics concerning moisture conditions and rabbit grazing. Relations between diversity and stability are demonstrated on the basis of yearly descriptions of one permanent transect.

References

Gimingham, C.H. 1975. An introduction to heathland ecology. Oliver & Boyd, Edinburgh, 124 pp.

Heukels-van Ooststroom. 1975. Flora van Nederland. 18e dr. Wolters-Noordhoff, Groningen, 913 pp.

Leeuwen, C.G. van. 1966. A relation theoretical approach to pattern and process in vegetation. Wentia 15: 25–46.

Londo, G. 1971. Patroon en proces in duinvalleivegetaties langs een gegraven meer in de Kennemerduinen (with a summary). Thesis Nijmegen, 279 pp.

Maarel, E. van der. 1966a. Dutch studies in coastal sand dune vegetation, especially in the Delta region. Wentia 15: 47–82.

Maarel, E. van der. 1966b. Over vegetatiestructuren, -relaties en -systemen (with a summary). Thesis Utrecht, 170 pp.

Maarel, E. van der. 1966. On the use of ordination models in phytosociology. Vegetatio 19: 21–46.

Maarel, E. van der. 1975a. Observations sur la structure et la dynamique de la végétation des dunes de Voorne. In: J.-M. Géhu (ed.). Colloques Phytosociologiques I, p. 167–183. Cramer, Vaduz.

Maarel, E. van der. 1975b. Small-scale changes in a dune grass-land complex. In: W. Schmidt (ed.) Sukzessionsforschung, p. 123–134. Cramer, Vaduz.

Maarel, E. van der. 1976. On the establishment of plant community boundaries. Ber. Deutsch. Bot. Ges. 89: 415–443.

Maarel, E. van der & C.G. van Leeuwen. 1967. Beziehungen zwischen Struktur und Dynamik in Ökosystemen. In R. Tüxen (ed.) Syndynamik. Ber. Int. Symp. Rinteln. (Still not published).

Maarel, E. van der & V. Westhoff. 1964. The vegetation of the dunes near Oostvoorne, The Netherlands. Wentia 12: 1–61.

MacArthur, R. 1955. Fluctuations of animal populations and a measure of community stability. Ecology 36: 533–536.

Margalef, R. 1958. Information theory in ecology. Gen. Syst. 3: 36–71.

Sloet van Oldruitenborgh, C.J.M. van. 1976. Duinstruwelen in het Deltagebied. (with a summary). Thesis Wageningen. Meded. Landbouwhogeschool Wageningen 76-8: 1–112.

Westhoff, V. 1965. Plantengemeenschappen. In: 'Het leven der planten', p. 288–349. De Haan, Zeist & van Loghum Slaterus, Arnhem.

Westhoff, V. & A.J. den Held. 1069. Plantengemeenschappen in Nederland. Thieme, Zutphen, 324 pp.

Westhoff, V. & E. van der Maarel. 1973. The Braun-Blanquet approach. In: R.H. Whittaker (ed.) Handbook of Vegetation Sciene, Part V, p. 617–726. Junk, The Hague

Williams, C.B. 1964. Patterns in the balance of nature. Academic Press, London-New York, VII + 324 pp.

SPATIAL AND TEMPORAL VARIATION IN THE VEGETATION OF DUNE SLACKS IN RELATION TO THE GROUND WATER RÉGIME[*,**]

Dick van der LAAN[***]

Institute for Ecological Research, Department of Dune Research, Oostvoorne

Keywords:
Dune slacks, Netherlands, Ordination, Vegetation succession, Voorne dunes, Water-table

Introduction

The aim of this study is to contribute to the understanding of the behaviour of plant species in a dynamic environment. Such a study provides information on changes in vegetation and, as environmental data were collected at the same time, suggestions can be made as to the causes of the changes which were recorded. It will be evident that in the environment under study, viz. the wet dune slack, where the vegetation is influenced by the proximity of the water-table, much attention is paid to the registration of the ground water régime (cf Ranwell 1959).

The data set consists of yearly registrations of the vegetation on permanent plots and additional analyses of soil and ground water conditions.

The dune slacks

The dunes on Voorne are part of the 'Younger dunes' (Jonge Duinlandschap) of the Dutch dunes (Faber 1960,

* Contribution to the Symposium of the Working Group on Succession Research on Permanent Plots held at Yerseke, the Netherlands, October 1975.

** Nomenclature of phanerogams follows Heukels & van Ooststroom 1975, Flora van Nederland, 18e druk Wolters-Noordhoff, Groningen; that of bryophytes Margadant 1959, Mossentabel, 3e druk, Amsterdam.

*** The author is very much indebted to Mr. P.A. Bakker, Society for the Preservation of Nature reserves in the Netherlands, for the use of the 1966 relevé of site 46, and to Mr. W. Smant for his assistance in field work and the preparation of the data.

Zagwijn 1971). According to the plant geographical system given by van Soest (1929, see also in Heukels – van Ooststroom, Flora van Nederland) the Netherlands are devided into eleven districts. The dune area covers two districts. The plant geographical border line between these two districts is near Bergen (Province of Noord-Holland). North of Bergen lies the 'Waddendistrict', to the south we find the 'Dune district'. The island of Voorne belongs to the Dune district which is, unlike the Waddendistrict, characterized by sandy soils rich in lime. The original $CaCO_3$-content of the sand of which the dunes of Voorne are built up, is 5–15%. See Van der Maarel & Westhoff (1964) for a general description of the area.

The sample sites which have been selected for this presentation are situated in different slacks. These slacks are relatively young. Concerning the sites the following points can be noted:

– site 46 and the transect lie in a primary dune slack, situated behind the present coastal ridge. Its development started in the beginning of the 20th century. About 1935 the backshore level became enclosed by *Ammophila* dunes. In 1953, after the storm disaster of January 31th, there was a break-through. During the period 1953-1975 sea water occasionally penetrated into the slack in winter. This situation ended definitively in 1975, when, according to the Delta project, the Dutch coastal defence scheme, a sand dam was constructed.

sites 73 and 140 are located in a primary dune slack, also situated behind the present coastal ridge. Its development started in 1926. About 1935 the backshore level became similarly enclosed by *Ammophila* dunes. Here, too, there was a break-through in 1953. In the same year the coastal ridge was closed again in a natural way, with some human help.

- site 3 is situated in a primary dune slack, which was formed between 1910 and 1926.
- site 57 is located in a secondary dune slack situated in the medieval dunes and blown out about 1930–1940.

The water-table

As is shown in Fig. 1 the water-table is markedly fluctuating. In winter it is high, from April-September it gradually falls. After that time it is rising again to the high winter level. Depending on precipitation and evaporation there are considerable differences between the years (Ranwell 1959). As will be shown below, the vegetation reacts to these year-to-year differences. However, since the water-table at a given moment is not representative for the ground water régime as a whole, it is necessary to have an overall parameter to compare the ground water régimes of different sites or of different years. From previous studies it appeared that the average water level during the vegetation period (April-September) is a suitable parameter (van der Laan 1970).

Spatial variation in the vegetation in relation to the water-table

The relation between vegetation and water-table is demonstrated by a study of the vegetation and some environmental factors in a transect (13 × 1.5 sq. meter) in a primary dune slack on Voorne (Fig. 2). In 1972, 13 relevés were made in this transect according to the Braun-Blanquet method. The height of site relative to the average ground water level was determined and analyses of soil samples

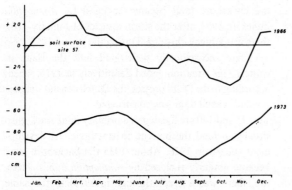

Fig. 1. Seasonal fluctuation of the ground water-table in a secondary dune slack on Voorne in two different years.

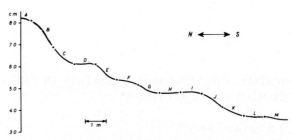

Fig. 2. A cross-section showing the height of site relative to the average water-table (April-October 1972) of a permanent transect in a primary dune slack on Voorne.

were carried out. The results of the soil analysis of the A-horizon are:

Thickness of the A-horizon		8–15 cm
Organic matter in the A-horizon	(% of dry matter)	2.3–4.5
Total nitrogen	(% of dry matter)	0.057–0.151
Calciumcarbonate	(% of dry matter)	5.0–9.0
Cl^-	(% of dry matter)	0.003–0.019
pH KCl		7.5–7.9
Volume weight (layer 0–5 cm, g/cm^3)		0.8–1.1

It appears that in spite of the minor spatial variation of the soil factors mentioned, the variation in vegetation is considerable. In Table 1 the relevés of the transect are are reproduced. To facilitate the characterization of the habitat the environmental factors involved are shown in Table 2. An ordination of the relevés of Table 1 was carried out according to Bray & Curtis (1957), based on presence and absence of the species and using the index of Czekanowski (Sörensen 1948). The position of the relevés on the X-axis, being the main line of variation, is shown in Figure 3. Because there was too little differentiation in the second and third dimension, the Y- and Z-axis have been left out of consideration.

Subsequently, the correlation between the respective environmental factors and the position of the relevés on the X-axis as obtained with the above method was calculated. The following correlation coefficients were found: ($n = 13$).

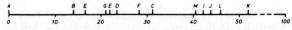

Fig. 3. Position on the X-axis obtained by ordination of 13 relevés of a permanent transect in a primary dune slack on Voorne. The Y-axis is left out of consideration.

Table 1. Vegetation analyses of the transect in a primary dune slack (1972)

Site	A	B	C	D	E	F	G	H	I	J	K	L	M
Herb layer % cover	20	85	80	75	80	85	80	80	80	30	30	40	45
Height in cm	10	10	10	5	10	13	10	10	10	2	2	3	2
Moss layer % cover	5	40	80	95	95	98	40	20	80	80	85	70	90
Number of species	29	35	33	31	35	37	37	31	31	34	29	33	33
Height relative to average water table (Apr.-Oct.)	84	72	57	51	52	55	64	64	52	45	41	39	40
Fragaria vesca	+	+											
Crepis capillaris	r	r											
Asparagus officinalis	r	r											
Brachytecium rutabulum	2m	2b											
Cirsium arvense	+	+		+	+								
Taraxacum species	+	+		+	+		+	+					
Carex flacca	r		2a	2a	2a	2a	1	1	1				
Holcus lanatus	+	2a	+	+	+	+	2b	2a	+	r			
Sonchus arvensis. maritima	r			+	r	+	+	+	+	+		+	
Hippophae rhamnoides	+				+		+	r				r	r
Pulicaria dysenterica	r	+	+	+		r						+	+
Rhamnus catharticus	r	r			r	r	r	r				r	r
Pyrola rotundifolia	+	+		r	r	+	+	+	+				r
Rubus caesius	+	1	+	r	r	r	+			r	r	r	
Betula pendula	+	r		r		r				r	r	r	
Calamagrostis epigeios	2m	2m	2m	1	2m	1	1	2m	2m	+	r		r
Salix repens	2a	2b	3a	3a	3b	4a	3b	3b	4a	+	+	+	+
Calliergonella cuspidata	+	+	3a	5b	5b	5b	3b	2b	4b	+	+	+	
Eupatorium cannabinum	+	+	+	1	1	+	1	1	1	1	1	2m	2m
Euphrasia officinalis	2m	2m	1	1	1	2m	1	2m	+	+			+
Galium uliginosum	+	1	2a	1	2b	2a	2a	2b	2a	+	+	+	+
Mentha aquatica	+	+	1	2m	2m	2a	+	+	+	2m	2a	2a	2a
Hydrocotile vulgaris	+	2b	2a	2a	2a	2b	2a	2m	2a	2m	1	1	1
Parnassia palustris	+	+	1	+	+	+	+	+	+	2a	2a	2b	2b
Juncus articulatus	+	+	1	2m	2m	+	+	+	1	1	1	2m	1
Centaurium littorale	r		+	r									
Agrostis stolonifera		+	1	+	1	+	+	+	1	2b	2b	2b	2a
Campylium polygamum		2b	3b		+	+	+	+	2a	4b	3b	3b	4b
Eleocharis palustris		+	1	1		+			+	1	2a	2a	2a
Riccardia species		+			+	+	+	+	+	+	+	+	2b
Galium palustre		+	1	+		+			+		r	r	
Cardamine palustre		+	1	+	+	1		r	+				
Sagina nodosa	r	r					+						
Bryum pseudotriquetrum		+				+	+						
Cirsium vulgare		+					r						
Poa pratensis		+	+			+							
Epilobium parviflorum	r	r											
Senecio jacobaea							+	+					
Festuca rubra					+	1	2m	2m	+				
Gentianella amarella			+	2a			+	+		r			
Epilobium palustre		r	r	r	r								
Salix cinerea		r		r	r	r							
Lycopus europaeus		+	+	+	+	r				r		+	
Prunella vulgaris		+		r	+	1	2b	1	+			+	+
Cirsium palustre		+	+		2a	1	+	+	+		r	+	r
Juncus gerardii			2m	2a	2a	2a	2m	2m	2m	2m	1	1	+
Glaux maritima				+	+	+			+	+	+	1	
Carex serotina. pulchellum							r	r	+	2a	2a	2a	2a
Liparis loeselii		r							r	r	+	+	+
Potentilla anserina									+	+	+	+	+
Drepanocladus species									+	2a	3b	3a	2b
Centunculus minimus										+	2m	+	2m
Samolus valerandi										+	+	+	1
Plantago major. pleiosperma												+	1

Addenda: site J: Ranunculus flammula +, site M: Centaurium pulchellum +.

Table 2. Analysis of the upper soil layer of the transect

Site	Organic matter (%)	A-horizon (cm)	N (total) (%)	CaCO$_3$ (%)	Cl$^-$ (%)	Bulk density (g/cm^3)	pH (KCl)
A	2.8	10	0.084	6.1	0.004	1.1	7.6
B	4.3	10	0.151	6.1	0.003	0.9	7.5
C	4.5	12	0.145	5.6	0.011	1.0	7.6
D	4.2	12	0.124	5.0	0.014	0.8	7.6
E	3.4	15	0.093	5.1	0.019	0.8	7.7
F	3.1	13	0.069	5.7	0.013	0.9	7.8
G	3.6	15	0.087	5.8	0.003	0.9	7.6
H	4.2	10	0.116	6.1	0.003	0.8	7.6
I	3.7	8	0.098	7.6	0.005	0.8	7.7
J	2.8	8	0.079	8.2	0.013	1.1	7.9
K	2.4	8	0.069	9.0	0.014	1.0	7.9
L	2.5	8	0.062	8.8	0.011	1.0	7.9
M	2.3	8	0.057	8.5	0.008	1.0	7.9

a. position of the relevés
on the X-axis and average water-table (April–September) in 1972
$r = 0.832$ P < 0.005
b.   ,,   ,, percentage of organic matter
$r = 0.542$ P < 0.025
c.   ,,   ,, thickness of the A-horizon
$r = 0.569$ P < 0.025
d.   ,,   ,, percentage CaCO$_3$
$r = 0.675$ P < 0.01
e.   ,,   ,, percentage nitrogen (N tot)
$r = 0.53$ (n.s.)
f.   ,,   ,, percentage Cl$^-$
$r = 0.306$ (n.s.)
g.   ,,   ,, pH KCl
$r = 0.309$ (n.s.)
h.   ,,   ,, volume weight
$r = 0.239$ (n.s.)

The average water-table in the growing season clearly shows the highest correlation. The other factors showing any level of significance are not independent of the water-table, as follows from the correlation coefficients shown in Table 3.

	water-table	organic matter	A-horizon	$^0_0$ CaCO$_3$
water-table	–	0.505 P < 0.05	0.313 n.s.	0.539 P < 0.05
organic matter		–	0.175 n.s.	0.617 P < 0.025
A-horizon			–	0.756 P < 0.005
$^0_0$ CaCO$_3$				–

Table 3. Correlation coefficients between environmental factors.

From previous studies it also appeared that the vegetational composition of transects in dune slacks was highly correlated with the average ground water level (Londo 1971, van der Laan 1974).

Temporal variation in the vegetation in relation to the water-table

Changes in the vegetation in the course of years have been investigated by means of yearly Braun-Blanquet relevés on permanent plots, in combination with environmental analyses. The effectiveness of such an approach will be illustrated using the results of a ten year study (1966–1975) on one permanent plot situated in the same dune slack as the transect described above. The relevés and the average water-table during the vegetation period in the respective years are shown in Table 4. The relevés were ordinated in the same way as those of the transect. In Fig. 4 their position along the X-axis is shown.

Treatment of vegetation data using such a multivariate method is an effective way to trace trends of succession (van der Maarel 1969, Londo 1971). If the development over the years is regular in one direction, the positions of the relevés along the X-axis of the ordination model will correspond to the successive years. When this trend appears to be disturbed, it may be an indication that an environmental change beyond the expected range has affected the vegetation. This is illustrated in the distribution of the series of relevés of site 46: the deviating position of the year 1975 may be related to the extreme high water-table in that year (Fig. 4). This supposition is confirmed by the repetition of this phenomenon which is found in a number of other permanent plots which were analysed

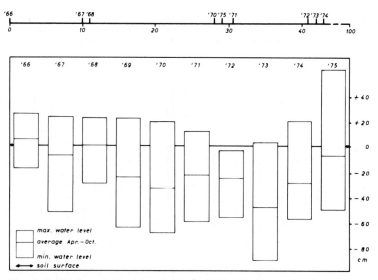

Fig. 4. Ordination of the series of relevés 1966–1975 of a permanent plot (site 46) and the seasonal ranges of the free water-table level relative to the soil surface in a primary dune slack on Voorne. In 1969 no relevé was made.

in the same way (Fig. 5). From this figure it appears that an exceptionally high average water-table during the vegetation period is not the only factor determining anomalous relevés positions on the X-axis of the respective ordinations. E.g. on site 57 the maximum water-table of the year 1973 was exceptionally low and the total range is much smaller than usual. On site 73 the ground water régime in 1969 deviates considerably from other years, in both the maximum and the minimum level as well as the total range. On site 140 the maximum water-table of 1972 was extremely low, besides, it is the only year during the observation period in which this site was not inundated. Apart from 1966 the ordination of site 3 did not show irregularities. This is connected with the exceptional character of the dune slack in which this plot is situated. In this slack there is an artificial control of the water level in spring; the surface water which is dammed up during autumn and winter and is flowed off each year in the beginning of March. This regulation reduces the fluctuation of the ground water-table considerably, resulting in a much more regular succession of the vegetation. The anomaly of 1966 is caused by the extreme high amount of precipitation in spring and summer which overruled the stabilizing effect of the yearly draining.

The interpretation of these results may be facilitated by the additional information that in general the average water-table during the vegetation period is the ground water régime parameter giving the best correlation of all the ground water parameters investigated. The anomalies are induced by extreme water-tables in exceptional years, e.g. a very long inundation period in spring, an extremely high maximum level or an extremely low minimum level. In conclusion, it can be stated that differences in the height of a site relative to the water-table affect the composition of the vegetation as such and abundance - cover values of the species (Tables 1 and 4).

The amount of variation, both in space and in time, appears to be related to the differences in ground water régime. An examination of the figures showing the ordination of the various permanent plots and the corresponding figures reflecting the water régime learns that the level of temporal variation is related to the height of site above the water-table (Londo 1971). This means that differences in height relative to the water-table may be of importance; the influence of a particular ground water régime can be differerent. Figure 6 shows an attempt to reproduce this conclusion. This figure reflects a hypothetical gradient divided in the zones A, B, C, D and E. The effect of the seasonal fluctuations and the differences of the water-table between one year and another on each of the various zones is supposed to be different.

The various zones might be characterized as follows:
zone A: permanently inundated
zone B: periodically flooded
zone C: never flooded, permanently under the direct

69

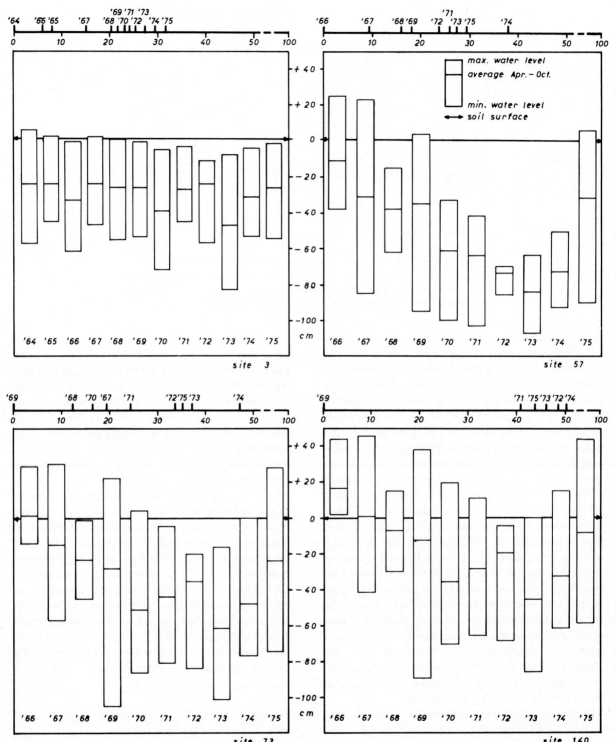

Fig. 5. Ordination of series of relevés of selected permanent plots (site 3, 57, 73 and 140) and the seasonal ranges of the free water-tabel level relative to the soil surface in dune slacks on Voorne.

Table 4. Vegetation analyses of permanent plot 46

Year of analysis	'66	'67	'68	'70	'71	'72	'73	'74	'75
Tall herb layer % cover	15	5	8	5	1	–	1	–	1
Height in cm	70	75	50	50	25		15		15
Low herb layer % cover	100	70	75	85	90	60	70	70	90
Height in cm	15	30	20	20	5	3	4	3	3
Moss layer % cover		10	25	98	95	90	100	98	95
Number of species	16	19	17	22	25	30	37	37	26
Height of site rei. average water table (Apr.-Oct.) in cm	+8	−9	−3	−26	−13	−12	−39	−22	+1
Tall herb layer:									
Scirpus maritimus	2b	2m	2a	2m	1				2m
Scirpus lacustris. glaucus	1	+	1	1	+				+
Mentha aquatica			1		r				
Juncus gerardii			+						
Salix repens					+		+		r
Parnassia palustris							+		
Phragmites australis									+
Low herb layer:									
Eleocharis palustris	2a	2a	2m	2m	2m	2m	2m	1	+
Juncus articulatus	1	2m	2m	1	2m	1	2m	2m	2m
Juncus gerardii	2a	1	2a	2b	2a	2a	2m	2a	2m
Mentha aquatica	1	2m	2b	4a	3b	3b	3a	2a	2a
Samolus valerandi	2b	2a	2m	+	+	1	1	1	+
Littorella uniflora	5	3a	3a	2a	2a	2a	2m	2m	4a
Galium palustre	1	1	1	2a	2m	+	1	2m	1
Agrostis stolonifera	1	2m	2b	3a	2b	2b	3a	3a	2a
Hydrocotile vulgare	1	2a	2b	2b	2b	3a	2b	2b	4b
Potentilla anserina	+	1	2a	2b	4a	3b	3a	2b	3a
Ranunculus repens	+	+	+	r	+	+	+	1	+
Gardamine palustris	+	+					+	r	
Lycopus europaeus	r				r	+	+	+	+
Chara species	2b								
Scirpus maritimus	+	+	2m	1	+	+	+	+	+
Eupatorium cannabinum		+		r	+	+	+	+	
Sonchus arvensis. maritimus		r			r	r	r	r	
Ranunculus flammula		r	r	r	+		+	r	r
Drepanocladus spec.		2a	+	1	1	1	1	2m	2m
Scirpus lacustris. glaucus			+	+	+	1	1	1	+
Salix repens			r	+	+	+	+	+	+
Campylium polygamum			2b	5b	5b	5b	5b	5b	5b
Phragmites australis				+	+	r	+	r	+
Epilobium palustre				r		+	r	r	
Plantago major. pleiosperma				r	r	+	+	+	+
Calliergonella cuspidata				+	+	+	+	2b	+
Hippophae rhamnoides					r	+	+	r	
Epipactis palustris					r	+	+	+	+
Taraxacum species						r	r	r	r
Carex serotina. pulchellum						+	+	+	+
Linum catharticum						r	+	+	
Pulicaria dysenterica						r	+	+	+
Parnassia palustris						+	2b	2b	
Cirsium palustre							r		
Rubus caesius							r	r	
Senecio jacobaea							r		
Cerastium holosteoides							r		
Euphrasia officinalis							r	+	
Solanum dulcamara								r	
Asparagus officinalis								r	
Centaurium littorale								0	
Rhamnus cathartica									r

71

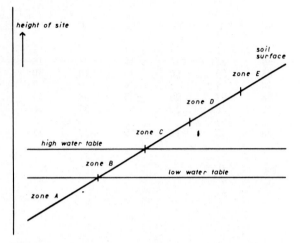

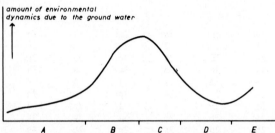

Fig. 6. Hypothetical gradient and postulated effect of ground water fluctuations on the vegetation related to the difference in height of site relative to the water-table.

influence of ground water, considerable fluctuations in water content of the soil.

zone D: never flooded, permanently under the direct influence of ground water, moderate fluctuations in water content of the soil.

zone E: never flooded, considerable fluctuations in water content of the soil, periodically beyond the direct influence of ground water.

Discussion

The preliminary results presented here clearly suggest that long term analysis of vegetation combined with measurement of fluctuating environmental factors may reveal novel information on both the population dynamics of individual species and the characteristic species composition of plant communities. It also appears that the decisive environmental factor causing floristic-sociological variation in vegetation may be a dynamical factor, in our case related to the average ground water-table in the vegetation period.

The scheme (Fig. 6) suggests how the environmental dynamics on a site may be related to the position of that site towards average maximum and minimum ground water level. It remains uncertain how to measure environmental dynamics (cf van Leeuwen 1966, van der Maarel 1976). The result of this study by no means fully prove the hypothesis put forward in this contribution. To explain the relation further, an elaboration of the approach as presented here is intended using more extensive material collected from permanent plots.

Summary

The aim of the study presented here is to determine the relation between the vegetation and the ground water régime in the dune slacks on the island of Voorne. A distinction is made between the spatial and the temporal variation in the vegetation.

On the basis of data from a transect the relation between the spatial variation and the water-table is determined. With respect to the temporal variation data recorded on permanent plots were used.

Both the variation in space and that in time proved to be related to differences in the ground water régime. The supposition is put forward that the amount of variation in time depends on the height of site relative to the water-table.

References

Bray, J.R. & Curtis, J.T. 1957. An ordination of the upland forest communities of Southern Wisconsin. Ecol. Monogr. 27: 325–349.

Faber, F.L. 1960. Geologie van Nederland, IV, Aanvullende hoofdstukken. Noorduyn, Gorinchem, 607 pp.

Heukels, H. & S.J. van Ooststroom, 1975. Flora van Nederland, Wolters-Noordhoff, Groningen, 913 pp.

Laan, D. van der. 1970. Vegetation research in the dune slacks on Voorne. Verh. Kon. Ned. Akad. Wetensch., afd. Natuurk., 2e Reeks, 59–61.

Laan, D. van der. 1974. Synecological research of the dune slacks on Voorne; Analysis of vegetational and environmental data. Verh. Kon. Ned. Akad. Wetensch., afd. Natuurk., 2e Reeks, 63: 93–96.

Leeuwen, C.G. van. 1966. A relation theoretical approach to pattern and process in vegetation. Wentia 15: 25–46.

Londo, G., 1971. Patroon en proces in duinvalleivegetaties langs een gegraven meer in de Kennemerduinen. Thesis, University of Nijmegen, 279 pp.

Maarel, E. van der. 1969. On the use of ordination models in phytosociology. Vegetatio 19: 21–46.

Maarel, E. van der. 1976. On the establishment of plant community boundaries. Ber. Deutsch. Bot. Ges. 89: 415–443.

Maarel, E. van der & V. Westhoff. 1964. The vegetation of the dunes near Oostvoorne, Netherlands (with a vegetation map). Wentia 12: 1–61.

Margadant, W.D. 1959. Mossentabel. 3e druk, Amsterdam, 155 pp.

Ranwell, D.S. 1959. Newborough Warren, Anglesey. I. The dune system and dune slack habitat. J. Ecol. 47: 571–601.

Soest, J.L. van. 1929. Plantengeografische districten in Nederland. De Levende Natuur, 33: 311–318.

Sörensen, T. 1948. A method of establishing groups of equal amplitude in plant sociology based on similarity of species content and its application to analyses of the vegetation on Danish commons. Biol. Skr. N.S. 5: 1–34.

Zagwijn, W.H. 1971. Vegetational history of the coastal dunes in the Western Netherlands. Acta Bot. Neerl. 20: 174–182.

VEGETATION DEVELOPMENT ON SALT-MARSH FLATS AFTER DISAPPEARANCE OF THE TIDAL FACTOR[*][**]

K. VAN NOORDWIJK-PUIJK[1], W.G. BEEFTINK[2] & P. HOGEWEG[3][***]

[1] Fazantenkamp 92, Maarssenbroek, The Netherlands
[2] Delta Institute for Hydrobiological Research, Yerseke, The Netherlands[****]
[3] Bioinformatica Group, Subfaculty of Biology, The University, Padualaan 8, Utrecht, The Netherlands

Keywords:
Cluster analysis, Colonization, Diversity, Spatial pattern, Succession, Temporal pattern, Vegetation development

Introduction

In the South-West Netherlands most sea-arms and estuaries of the rivers Rhine, Meuse and Scheldt have been cut off from the sea by a barrage. One of these former sea-arms, the present Lake Veere, was closed in April, 1961. As a result of this engineering work the tidal influence abruptly disappeared and from 1961 the water level is kept at Dutch Ordnance Level (NAP) in summer and at minus 70 cm NAP in winter. A few years after the barrage construction the salinity of the lake dropped to brackish values (12–8 $^0/_{00}$ Cl$^-$).

This tideless situation lead to desiccation, desalinization, and increased aeration of the soil, as well as to an increased rate of mineralization of organic matter in the emerging parts of the former tidal flats (Beeftink et al., 1971). The original marine biota died off within a period of some weeks (zoobenthos, phytobenthos, e.g. Zostera) to about one year (Spartina). The newly developing vegetation started in spring 1961, with thin and very local populations of Salicornia and Suaeda emerging from seed still available, and dispersed by tidal action before April, 1961.

In 1962 and 1963, 36 permanent plots were established

on the Middelplaten, a system of sand flats separated by channels and mainly consisting of three parts called Peninsula (130 ha), Big Island (32 ha) and Small Island (12 ha) (Fig. 1). The colonizing vegetation has since been examined by analysing these permanent plots yearly the result of which is called a relevé. In this paper some results of a numerical analysis of the complete data set over 12 years is presented. A more detailed discussion is given of the results obtained from the Big Island. Other details not included in this paper can be found in Van Noordwijk-Puijk (1976), while a further numerical analysis of some aspects is reported on by Hogeweg (1976).

Material and methods

The permanent plots were mainly located for distributional and geomorphological reasons because vegetation was very sparse or none at all at that time. Plot sizes vary from 7 to 120 sq. m, approximately twice the qualitative minimum area as appeared from vegetation analyses carried out later on.

Cover-abundance of all species of higher plants was estimated with the scale of Doing Kraft (1954). For the numerical analysis the Doing Kraft symbols were adapted. The cover estimate (01–10) for species covering more than 5% was changed to 10–100 percentage coverage. Below 5% coverage, where this scale indicates the number of individuals with r, p, a, m symbols, these scores were transformed to 1, 2, 3, 4% coverage respectively. The logarithms (zero corrected, i.e. if $x \geq 1$ then $x = \log x + 1$, else $x = x$) of these transformed values were used.

Cluster analysis and principal component analysis were

* Nomenclature of vascular plants follows Heukels-van Ooststroom, Flora van Nederland, 19th Ed. Wolters-Noordhoff, Groningen, 1977.
** Contribution to the Symposium of the Working Group for Succession Research on Permanent Plots, held at Yerseke, The Netherlands, October 1975.
*** The authors would like to thank Dr. K. F. Vaas and Dr. A.H.L. Huiskes (Yerseke) for reviewing the English text, and Mr. M.C. Daane and Mr. W. de Munck (Yerseke) for their help in producing the data.
**** Communication Nr. 172.

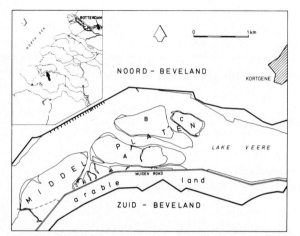

Fig. 1. The Middelplaten sand flats in Lake Veere. Situation of the permanent plots. A. Peninsula, B. Big Island, C. Small Island.

carried out using BIOPAT, Program System for Biological Pattern Analysis (Hogeweg & Hesper, 1972). Here we report the results of the cluster analysis only. Agglomerative clustering techniques were used, with minimal increase of the mean sum of squares as clustering criterion (Ward 1963). This criterion was calculated iteratively as proposed by Wishart (1969):

$$D_{lk} = \frac{(N_i + N_k)}{N_i + N_j + N_k} D_{ik} + \frac{N_j + N_k}{N_i + N_j + N_k} D_{jk} - \frac{N_k}{N_i + N_j + N_k} D_{ij}$$

where D_{lk} is the 'distance' between cluster l and k; l designates the newly formed cluster by combining cluster i and cluster j; N_i is the number of elements in cluster i. At each step the two clusters with the smallest 'distance' are combined. The process is started with each relevé as a seperate cluster and the mean square distance as distance criterion, i.e.

$$D_{ij} = \frac{1}{m} \sum_{k=1}^{m} (V_{ik} - V_{jk})^2$$

where m is the number of characters (plant species), and V_{ik} designates the abundance of the kth species in the ith relevé, expressed in the transformed logarithmic scale value.

The result of the cluster analysis is a dendrogram indicating the hierarchical similarity structure of the data set. From the dendrogram clusters were extracted using the optimality criterion of Beale (reported by Kendall 1972). Two levels of partitioning were recognized, representing a coarser and finer grained classification of the data set.

The clusters were described in terms of the mean and

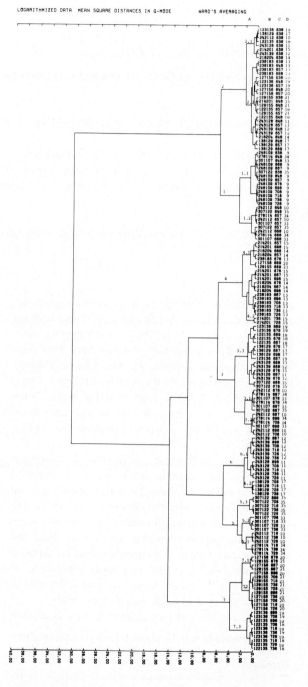

Fig. 2. Dendrogram of the 1963–73 relevés in the Big Island. The 7 resp. 17 clusters have been classified at two levels according to the criterion of Beale, reported by Kendall (1972). A = numbers of clusters, B = co-ordinater, C = date, year and month, D = numbers of permanent plots.

76

variance of the species coverage in the cluster and in terms of the frequency of occurrence of the species. Moreover, the 'importance' of the species in distinguishing the clusters from each other was determined by using as index Kruskal-Wallis' one way analysis of variance (Hogeweg 1976).

Results

The Big Island

The dendrogram is partitioned at two levels yielding 7 and 17 clusters respectively, according to the criterion of Beale. These clusters are designated by a figure of 2 digits indicating the two levels of classification. The dendrogram is shown in Fig. 2.

Major variation in the vegetation

For each of the 7 main clusters we shall briefly discuss the species groups present in the relevés composing the clusters, as well as their occurrence in space and time (see Table 1).

Cluster 1 is composed of stands which occurred in the first years (1963–66). An exception is quadrat 9 of which all relevés belong to this cluster. The vegetation is poor in species and coverage is low, not exceeding 50%. *Aster tripolium*, *Spergularia marina*, *Puccinellia fasciculata*, *P. distans* and *Salicornia europaea*, all halophytes, are represented in more than 50% of the relevés. In most relevés one or two of these species are dominant (Table 1).

Cluster 2 is also composed of stands developed in the first years. In most relevés the number of species is higher than in cluster 1. Coverage is generally low. *Matricaria inodora*, *Poa annua*, *Atriplex hastata* and *Puccinellia distans* are found in all relevés. *Matricaria* and *Senecio vulgaris* are the most abundant species. *Atriplex littoralis*, *Suaeda maritima*, *Polygonum aviculare* and *Senecio vulgaris* developed optimally. Nitrophilous plants have a great share in this cluster.

The relevés of cluster 3 date from 1966–70. They are characterized by dominance of *Epilobium* species: *E. adnatum*, *E. hirsutum*, *E. parviflorum*, *E. adenocaulon*. The mean coverage of *Epilobium* seedlings is 20%. Besides these species *Pao annua* and *P. trivialis* are always present with a relatively high mean coverage of 20% and 15% respectively.

Cluster 4 resembles cluster 3 in several ways. The relevés of cluster 4 extend from 1965 to 1973. *Epilobium* species are dominant as well, but here particularly *E.*

adenocaulon and *E. (Chamaenerion) angustifolium*. Cluster 4 can be considered as a poor variant of cluster 3, poor in species as well as in total coverage. *Erigeron canadensis* is the most typical species for this cluster.

Relevés of cluster 5 are found in the years 1969–73. *Poa trivialis* and *P. pratensis* are dominant species (mean coverage about 60%), sometimes in combination with *Scirpus maritimus* and *Epilobium hirsutum*. *Scirpus maritimus* and *Carex otrubae* are typical for this group.

The relevés of cluster 6 date from 1968–73. *Poa trivialis* and also *P. pratensis* are dominant. The mean coverage of *P. pratensis* has increased and that of *P. trivialis* has decreased with respect to cluster 5. The total number of species per relevé is relatively high (about 40) in this cluster. Among the typical species *Fragaria vesca*, *Gnaphalium luteo-album*, *Sagina procumbens*, *Trifolium repens*, *Bellis perennis* and *Holcus lanatus* can be mentioned.

The relevés of cluster 7 date from 1967–73. The number of species in these relevés is much lower (10–15) than in the clusters 5 and 6. There is always one highly dominant species, mostly *Calamagrostis epigejos*, sometimes *Festuca rubra* or *Epilobium hirsutum*.

Table 1. Species distribution in the main clusters of the Big Island. Only those species having a Kruskal-Wallis index > 10 are induced.

Cluster Nr.	Presence (%) 1	2	3	4	5	6	7	Mean coverage (%) 1	2	3	4	5	6	7
Salicornia europaea	57	19						5	3					
Puccinellia fasciculata	74	23	7					12	2	1				
Spergularia media	43	45	4					2	1	2				
Spergularia marina	91	74	32					10	4	2				
Puccinellia distans	65	94	43	48	13			9	5	3	3	3		
Atriplex hastata	39	97	7	29				2	5	2	2			
Atriplex littoralis	4	61						1	3					
Matricaria chamomilla	30	65	43	10				2	2	1	2			
Polygonum aviculare	17	74	14	19				1	2	1	2			
Sonchus oleraceus		65	7	38		7	8		2	1	3		1	1
Suaeda maritima	30	48		7				3	4		3			
Senecio vulgaris	9	81	61	14		7		2	5	2	1		2	
Juncus bufonius	39	52	79	14	27	40		18	5	4	1	3	2	
Capsella bursa-pastoris		39	7	24					2	1	1			
Aster tripolium	96	81	93	38	80	47	13	10	2	5	2	4	1	1
Matricaria inodora	30	100	100	100	67	100	13	2	14	5	3	7	3	2
Poa annua	30	100	100	95	73	93	38	3	6	20	7	5	3	5
Epilobium parviflorum	17	55	100	100	87	100	38	4	2	10	4	3	3	4
Plantago coronopus	13	10	71	71	87	60	4	1	1	2	2	3	3	1
Plantago major	4	35	93	90	100	87	33	1	1	3	2	4	3	2
Poa trivialis	9	10	96	67	100	100	79	2	1	15	4	45	35	4
Poa pratensis	17	52	93	86	100	100	100	1	2	5	3	20	29	10
Taraxacum officinale		58	79	95	93	100	92		1	2	2	2	5	2
Cirsium arvense		26	57	95	67	100	100		1	2	4	3	5	7
Cirsium vulgare		19	68	100	67	100	54		1	2	2	3	4	3
Cerastium holosteoides	9	6	89	71	93	100	42	2	1	5	2	3	4	1
Epilobium hirsutum		10	89	71	100	100	100		1	2	6	3	12	
Juncus gerardii	17	45	68	48	67	93	50	2	1	2	2	6	2	3
Epilobium adnatum	22	58	100	90	40	27	13	2	2	6	8	1	2	2
Sagina maritima	39	6	82	43	27	33	13	3	6	6	3	2	3	2
Epilobium seedlings			43	24	13	13	13			20	7	4	4	3
Ranunculus sceleratus	9	57						1	2					
Epilobium adenocaulon		23	86	71	40	27	38		2	11	10	2	5	8
Sonchus asper	32	68	52	7	47	13		1	1	1	1	2	2	
Centaurium pulchellum	4	3	75	5	53	33	8	2	2	3	1	3	2	3
Erigeron canadensis		39	54	100		60	8		2	2	5		3	3
Chamaenerion angustifolium		48	36	100	7	100	58		2	2	10	1	4	4
Agrostis stolonifera		3	18	52	87	73			1	1	4	2	1	
Juncus articulatus	13		54		100	80	4	1		3		5	2	1
Carex otrubae	4	6	29	48	73			1	1	2	2	4		
Scirpus maritimus	13		21	19	87			2		3	2	12		
Tussilago farfara		16	21	5	47	40			2	2	1	1	3	
Ranunculus repens			29		93	87	13			1		3	2	1
Centaurium vulgare			25	38	60	93	8			1	2	2	3	1
Bellis perennis			68	10	73	100	38			2	1	3	6	4
Sonchus arvense	6	36	81	60	80	79		2	1	2	1	2	1	
Crepis capillaris			36	67	33	93	8			1	2	1	2	1
Gnaphalium luteo-album		3	29	57	100		4		1	1	2	2		1
Sagina procumbens			32	71	27	100	17			2	6	2	3	1
Trifolium repens			32	33	53	100	21			1	1	4	3	1
Salix repens			11		47	47	17			1		1	1	1
Ranunculus sardous		3	29		47	53	8		1	1		1	2	2
Dactylus glomerata	4	3	14	5	27	73		1	1	1	1	1	2	
Fragaria vesca			25	19	13	80	13			1	1	1	1	1
Holcus lanatus			4		7	60				1		1	2	
Calamagrostis epigejos	13	25	29	53	87	100		1	10	2	3	3	40	
Festuca rubra	13	14	10	20	20	58		2	1	45	1	1	39	

77

Detailed pattern of variation

A more detailed picture can be obtained by dividing the data set into 17 subclusters distinguished at the second level of classification. In the rest of this paper we shall concentrate on this level of classification.

The subclusters are characterized by more subtle floristic differences as compared to the main clusters described above. Cluster 1.2 is distinguished from cluster 1.1 by a slightly increased coverage of the vegetation including the establishment of *Juncus bufonius, Puccinellia distans, Poa annua* and *Matricaria* spp. Cluster 2.1 differs from its counterpart nr. 2.2 by the presence of halophytes (*Spergularia marina, Suaeda maritima*) and nitrophilous species (*Atriplex hastata, A. littoralis*), while cluster 2.2 shows a higher abundance in *Matricaria* species, *Puccinellia distans* and *Poa annua.*

Cluster 3 varies in different ways: Both 3.1 and 3.2 are rich in *Epilobium parviflorum, E. adenocaulon* and *E. adnatum*, but 3.1 has a co-dominance of *Aster tripolium.* Cluster 3.2 is rich in *Epilobium* seedlings instead of *E. adnatum*, and shows poor growth in *Matricaria.* Cluster 3.3 resembles 3.2 but *Poa annua, P. trivialis* and *Matricaria inodora* dominate more.

In cluster 4.1 nitrophilous species have retreated, and *Poa annua* and *Epilobium adnatum* are mostly dominant. Cluster 4.2 is marked by dominating *Chamaenerion angustifolium* and co-dominating *Epilobium adenocaulon* and *Poa trivialis*, while *Sagina procumbens* and *Erigeron canadensis* are very abundant.

Cluster 5 is more variable: 5.1 differs from its counterparts by many individuals of *Juncus articulatus* and *Carex otrubae;* 5.2 by many *Plantago major* s.l., *Epilobium hirsutum, Scirpus maritimus* and *Eleocharis uniglumis*, and few *Poa pratensis;* 5.3 by many *Poa pratensis, Agrostis stolonifera* and *Trifolium repens.*

Cluster 6.2 differs from 6.1 by the abundance of *Sagina maritima, Plantago coronopus, Centaurium pulchellum, Tussilago farfara*, and the presence of species such as *Luzula campestris, Rumex crispus* and *Myosotis caespitosa.*

In cluster 7.1 especially *Festuca rubra* has developed (with co-dominance of *Epilobium adenocaulon*). Those species are for the greater part replaced by *Calamagrostis epigejos* and *Epilobium hirsutum* in cluster 7.2. Cluster 7.3 is distinguished from both others by a close vegetation of *Calamagrostis epigejos* and *Poa pratensis*, poor in other species.

Distribution patterns of clusters and their relation with some environmental factors.

Mapping the subclusters for the successive years gives an impression of their spatial relationships through time (Fig. 3). For that purpose, it has been assumed that the relevés are representative, not only for the vegetation inside the plots, but also for a certain area outside. These areas were schematically delimited with the aid of a vegetation map based on interpretation of false-colour air photos made in June 1973. It is further assumed that the areas represented by a plot were about the same in all years. Starting from these assumptions a picture has been made indicating the spatial distribution of the clusters through the years. Some parts of the Big Island are left white, because in 1973 they had other vegetation types than the nearest plot. In 1963 the representative areas for plot 33 and 34 are white, because no vegetation had established there at that time.

In 1963 and 1964 the plots in the SE part of the island belong to cluster 1 (first 1.1, later 1.2). All other plots, scattered over the rest of the island, are classified into cluster 2 (2.1, later 2.2). These two pioneer stages reflect environmental differences.

The SE part is lower (plots 0–20 cm above NAP level) and has a higher percentage of clay (14–25 % particles < 16 μm) than the remainder of the island (plots 32–76 cm above NAP level, clay 4–15 %). The clay percentage and the height above the water level are correlated with desalinization rate: The higher and more sandy parts of the island showed a more rapid desalinization than the lower and more silty parts (see also Beeftink et al. 1971). This is reflected by vegetational colonization: cluster 1 was found on slowly desalinizing soils and cluster 2 on more rapid desalinizing ones.

Around plot 13 desalinization was most rapid, already in 1963 dropping down to less than $1^0/_{00}$ chloride. This point lies 76 cm above the NAP level and has a low percentage of clay. These conditions cause the vegetation to bypass cluster 2.1 starting with cluster 2.2 in 1963. Similar processes could be present in the plots 20 and 21.

In 1965 the vegetation at the N edge of the island entered cluster 4.1. Contrary to the other shores which are much lower and more gradual, this edge rises up 55–60 cm from the water like a microcliff. The vegetation characterized by cluster 4.1 spread afterwards over the whole higher E half of the island. Next the vegetation of cluster 4.2 spread over the area in the same manner. In the more clayey zone, originally occupied by cluster 1, the clusters 3.1, 3.2 and 5

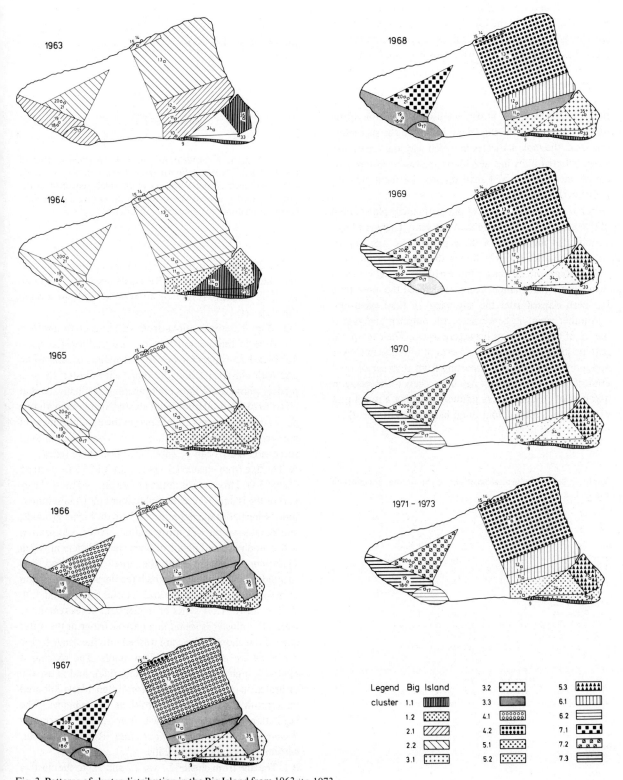

Fig. 3. Patterns of cluster distribution in the Big Island from 1963 |to 1973.

developed successively. Between these two zones there is a transitional one where cluster 2 is succeeded by 3.3 and 6.1. In the W part of the island the clayey layers alternate with the sandy soil and there, succession runs from cluster 2 via the clusters 3.3 and 4.1 towards 6.2 and 7.1–3.

Successional lines

In Table 2 is indicated to which cluster the relevés of the permanent plots of the Big Island belong in the successive years, i.e. the table indicates in which way the vegetation of the selected plots has developed, expressed in terms of cluster membership. We shall discuss the most striking points in this development.

In most cases the relevés of one permanent plot remain in the same cluster for more than one year. There is a large variation in the length of those periods, i.e. the rate of succession varies strongly among the plots. But the later in the succession, the longer those periods are. This means that vegetational changes decrease in speed the more time has been elapsed after the 'big bang' of tidal exclusion.

Another point is concerned with number and distribution of the transitions between clusters. Their temporal pattern suggests a wave-like process of alternating slower and more rapid succession starting from the year of tidal exclusion. In spite of this wave-like process the increase in numbers of clusters is very gradual, reaching a maximum value 10 years after colonization has started (Table 2).

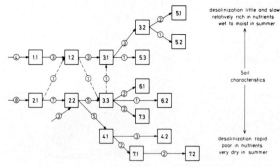

Fig. 4. Diagram of cluster transitions in the permanent plots of the Big Island. Numbers in arrows indicate numbers of transitions. For each cluster there are plots which remained in the same cluster for several years. Arrows indicating these within-cluster transitions are omitted in this figure.

The direction of transitions from one cluster to the next was the same for all plots: Reversals never occurred (see also Fig. 4). There is no indication of any retrogression in this data set.

In Fig. 4 the transitions from one cluster to another, discussed in the previous section, are outlined (compare Table 2). There appear to be two evident lines of succession, one with cluster 1.1 and the other with cluster 2.1 as a pioneer phase. Both lines are branching gradually, so that up to 1970 the total number of clusters increased with one per year. This process culminated in the eastablishment of nine terminal pathways in succession. These main lines of succession (A, B) can be subdivided into three series:

A. The line from cluster 1.1 via 1.2, 3.1 and 3.2 to cluster 5 (Table 3A). This series is limited to the SE and most clayey part of the island, where the soil remains moist in summertime owing to drainage conditions from the higher grounds, and desalinization was most gradual. After colonization with halophytes and plants of open stands rich in nitrogen *Epilobium* species invaded the area. This period was succeeded by a third one in which the floristic composition diverged in the plots in this area (cluster 5.1, 5.2 and 5.3). Within cluster 1.1 a series of sequential relevés remaining within this cluster is found in a narrow fringe at the water edge of flat shores. Algae are washed into the fringe by the waves or are growing in shallow pools. The soil has a relatively high salinity (4–50 ‰ chloride), and is usually saturated in summer and therefore often black-coloured underneath owing to concentrations of sulphides. The vegetation changed very slowly from mainly *Puccinellia distans* to dominating *P. fasciculata*. Recently, *Scirpus maritimus* is establishing locally.

B.1. The line from cluster 2.1 via 2.2 and 3.3 to the clusters 6.1, 6.2 and 7.3 (Table 3B). This succession series developed

Table 2. Table of cluster transitions in the permanent plots of Big Island.

Plot	63	64	65	66	67	68	69	70	71	72	73	Numbers of clusters
9	1.1	1.1	1.1	1.1	1.1	1.1	1.1	1.1	1.1	1.1	1.1	1
33		1.1	1.2	1.2	3.1	3.1	3.2	5.2	5.2	5.2	5.2	5
35	1.1	1.2	1.2	3.3	3.3	5.3	5.3	5.3	5.3	5.3	5.3	5
34		1.1	1.2	1.2	3.1	3.1	3.2	3.2	5.1	5.1	5.1	5
10	2.1	1.2	1.2	1.2	3.1	3.2	3.2	3.2	5.1	5.1	5.1	5
17	2.1	2.2	2.2	2.2	3.3	3.3	3.3	6.2	6.2	6.2	6.2	4
11	2.1	2.2	2.2	3.3	3.3	3.3	6.1	6.1	6.1	6.1	6.1	4
12	2.1	2.2	2.2	3.3	3.3	6.1	6.1	6.1	6.1	6.1	6.1	4
18	2.1	2.2	2.2	3.3	3.3	3.3	7.3	7.3	7.3	7.3	7.3	4
19	2.1	2.2	2.2	3.3	3.3	3.3	7.3	7.3	7.3	7.3	7.3	4
15	2.1	2.2	4.1	4.1	4.2	4.2	4.2			4.2	4.2	4
14	2.1	2.2	4.1	4.1	4.2	4.2	4.2					4
13	2.2	2.2	2.2	2.2	4.1	4.2	4.2	4.2	4.2	4.2	4.2	3
20	2.2	2.2	2.2	4.1	7.1	7.1	7.2	7.2	7.2	7.2	7.2	4
21	2.2	2.2	2.2	4.1	7.1	7.1	7.2	7.2	7.2	7.2	7.2	4
Numbers of transitions	9	4	7	9	4	8	2	2	0	0		
Numbers of clusters	3	3	4	5	6	7	8	9	9	9	9	

80

Table 3. (A-C). Species composition of the successional lines found in the permanent plots of the Big Island. Minimal and maximal coverages of species when presence is $\geq 75\%$. When presence is $< 75\%$ the symbol + is used.

Cluster Nr.	1.1	1.2	3.1	3.2	5.1	5.2	5.3
Puccinellia fasciculata	0-30	+		+			
Spergularia marina	0-40	3-40	+	+			
Aster tripolium	0-20	3-50	4-30	4	2- 4	0-10	
Puccinellia distans	+	3-25	+	+		+	
Matricaria chamomilla		0- 4	+	+			
Atriplex hastata	+	0- 3		+			
Juncus bufonius		4-50	0-10	4-10	+	+	
Matricaria inodora		0- 3	1- 2	2- 4	1- 2	+	+
Poa annua		0-10	4	4-20	2-10	0-20	+
Epilobium adnatum		+	4-10	2- 4	+	+	
Epilobium adenocaulon			4-30	4-20	+	+	
Ranunculus sceleratus		+	0- 3	2- 4		+	
Sagina maritima	+	+	0-30	3-10	+	+	
Centaurium pulchellum		+	0- 4	2- 4	0- 4	+	+
Epilobium parviflorum		+	10-30	4-20	2- 4	0- 4	0- 4
Cerastium holosteoides		+	2-10	2-20	2- 4	0- 4	1- 4
Juncus articulatus		+	2- 3	1- 5	4-10	2- 4	1- 2
Plantago coronopus	+	+	1- 3	2- 4	2- 4	1- 3	+
Plantago major		+	1- 3	2- 4	3- 4	3-10	1- 3
Epilobium hirsutum			0- 3	1- 4	3- 4	3-20	3- 4
Epilobium seedlings			+	4-50			+
Triglochin maritima	+	+	+	0- 3	+	+	
Poa trivialis		+	+	5-40	10-60	60-70	10-95
Poa pratensis		+	+	1-10	10-20	2- 3	10-60
Bellis perennis			+	1- 3	3- 4	2- 4	+
Cirsium vulgare		+	+	0- 3	2- 4	1- 2	
Carex otrubae		+	+	0- 3	1-10	2- 4	
Scirpus maritimus	+	+	+	+	0-10	20-40	0-10
Juncus gerardii	+	+	+	+	1-10	4-10	
Ranunculus repens			+	+	0- 3	4-10	2- 3
Cirsium arvense				+	2- 4	2- 3	
Taraxacum officinale		+		+	2- 4	1- 3	0- 1
Centaurium vulgare				+	1- 4	+	+
Agrostis stolonifera		+		+	1- 2	+	1-10
Tussilago farfara				+	1- 2	+	
Eleocharis pal.ssp.unigl.				+		10	
Sonchus arvensis		+	+		+	0- 2	+
Crepis capillaris		+	+		+	0- 1	
Dactylus glomerata			+		+	1- 2	
Calamagrostis epigejos					+	1- 5	
Solanum dulcamarum						0- 1	
Geranium dissectum			+	+		0- 2	
Eleocharis pal.ssp.pal.				+			0- 4
Salix repens			+		+		0- 1
Trifolium repens			+		+		1-10
Ranunculus sardous			+		+	+	0- 1

```
1.1 ──► 1.2 ──► 3.1 ──► 3.2 ──► 5.1
                  ╲              ╲──► 5.2
                   ╲
                   3.3            ──► 5.3
```

A

Cluster Nr.	2.1	2.2	3.3	6.1	6.2	7.3
Spergularia marina	2-20	+	+			
Juncus bufonius	0-10	+	+	+	+	
Suaeda maritima	1-20	+				
Atriplex littoralis	2- 4	+				
Polygonum aviculare	0- 3	+	+			
Atriplex hastata	3-20	0-10	+			
Senecio vulgaris	0-30	0-10	+	+		
Puccinellia distans	0- 3	1-20	+			
Aster tripolium	0- 2	0- 3	0-10	+	+	+
Matricaria inodora	1-30	3-60	2-30	1- 4	2- 4	+
Poa annua	1- 3	1-20	3-70	0- 4	4-10	+
Matricaria chamomilla	+	0- 3	+			
Epilobium adnatum	+	0- 4	2-30	+		+
Taraxacum officinale	+	0- 2	1- 3	3-10	2- 4	2- 3
Epilobium adenocaulon	+	+	0-30	+		+
Sonchus asper	+	+	1- 3	+		
Erigeron canadensis	+	+	0- 4	+	+	
Sagina maritima	+	+	0- 4	+	0- 4	+
Epilobium hirsutum	+	+	0- 4	2- 5	1- 2	2- 4
Poa trivialis	+	+	1-80	4-80	4-30	0-20
Poa pratensis	+	+	2-10	10-40	30-50	3-50
Bellis perennis		+	0- 4	3-10	4-10	0-10
Cirsium vulgare	+	+	0- 4	2-10	4	0- 4
Cirsium arvense	+	+	0- 4	3-10	3- 4	1- 4
Epilobium parviflorum	+	+	4-30	1- 4	1- 4	+
Cerastium holosteoides	+	+	0-20	2-10	4	+
Plantago major	+	+	0- 4	0- 4	+	
Calamagrostis epigejos	+	+	+	0- 4	1- 4	20-90
Juncus gerardii	+	+	+	1- 4	0- 2	2- 4
Juncus articulatus		+	+	0- 2	1- 3	+
Chamaenerion angustifolium	+	+	+	4-10	2	
Sonchus arvensis	+	+	+	0- 2	2- 3	+
Sagina procumbens	+	+	+	2- 4	2- 4	+
Gnaphalium luteo-album	+	+	+	1- 3	1- 4	+
Crepis capillaris	+	+	+	1- 4	0- 2	+
Centaurium vulgare	+	+	+	0- 4	4	+
Ranunculus repens	+	+	+	0- 3	1- 3	+
Trifolium repens	+	+	+	1-10	1- 2	+
Fragaria vesca		+	+	0- 2	2	+
Agrostis stolonifera	+	+	+	0- 2	+	
Dactylus glomerata	+		+	1- 3		
Holcus lanatus			+	0- 3		
Plantago coronopus	+	+	+	+	3- 4	+
Centaurium pulchellum	+	+	+	+	0- 4	+
Tussilago farfara	+	+	+	+	3- 4	
Geranium dissectum			+	+	0- 1	
Ranunculus sardous	+	+	+	+	0- 1	+
Phragmites communis	+	+	+		2	
Luzula campestris			+		1	
Rumex crispus	+	+	+		0- 2	0- 2
Myosotis caespitosa					0- 2	
Plantago lanceolata					0- 1	
Rubus spec.					0- 1	+

```
                          6.1
                          ╱
2.1 ──► 2.2 ──► 3.3 ──►── 6.2
                      ╲
                       ╲──► 7.3
```

B

Cluster Nr.	2.1	2.2	4.1	4.2	7.1	7.2
Spergularia marina	2-20	+				
Juncus bufonius	0-10	+	+	+		
Suaeda maritima	1-20	+				
Atriplex littoralis	2- 4	+				
Polygonum aviculare	0- 3	+	+	+		
Atriplex hastata	3-20	0-10	+	+		
Senecio vulgaris	0-30	0-10	+	+		
Matricaria chamomilla	+	0- 3	+			
Puccinellia distans	0- 3	1-20	1-20	+		
Aster tripolium	0- 2	0- 3	0- 4	+	+	
Matricaria inodora	1-30	3-60	2-10	1- 4	+	
Poa annua	1- 3	1-20	3-50	0-10	0-10	
Epilobium adnatum	+	0- 4	3-60	0-10	+	
Taraxacum officinale	+	0- 2	0- 2	2- 3	1- 2	0- 3
Chamaenerion angustifolium	+	+	1- 4	4-70	2- 4	3-10
Cirsium arvense	+	+	1-10	2- 4	2-10	4-20
Epilobium hirsutum	+	+	0- 2	0- 4	1-10	1-70
Epilobium parviflorum	+	+	3-10	2- 4	1-20	+
Plantago major	+	+	+	1- 3	0- 4	0- 2
Cirsium vulgare	+	+	1- 3	1- 4	0-10	+
Erigeron canadensis	+	+	4	2-20	+	
Juncus gerardii	+	+	0- 3	+	+	
Epilobium adenocaulon	+	+	+	0-30	4-50	+
Sonchus arvensis	+	+	+	0- 2	3- 4	3-10
Poa trivialis	+	+	+	0-30	+	0-10
Cerastium holosteoides	+	+	+	1- 4	+	+
Plantago coronopus	+	+	+	0- 3		
Sagina procumbens	+	+	+	0-20	+	
Gnaphalium luteo-album	+	+	+	0- 4		
Crepis capillaris	+	+	+	0- 4		
Sonchus asper	+	+	+	+	0- 2	
Calamagrostis epigejos	+	+	+	+	2-20	2-60
Festuca rubra	+	+	+	+	20-95	0-90
Hippophae rhamnoides	+	+	+	+	1- 2	+

```
2.1 ──► 2.2 ──► 4.1 ──► 4.2
                  ╲
                   ╲──► 7.1 ──► 7.2
```

C

on soils with somewhat lower clay contents running from the SW side of the island eastward along those of the former group. Desalinization proceeding at the soil surface is uncommon. The period governed by *Epilobium* species and *Poa annua* (cluster 3.3) has been preceded by a period in which nitrophilous and halophilous species were prevalent. It has been succeeded by a period in which a variety of vegetation types in which either a rather open vegetation with various herbs and perennial *Poa* spp. as dominant grasses (cluster 6), or a much more closed *Calamagrostis epigejos-Poa pratensis* assemblage is involved.

B.2. The line from cluster 2 via 4.1 to the clusters 7.1 and 7.2 or 4.2 (Table 3C). This sequence is found on the higher parts of the island where desalinization went very fast and the upper soil layers can be very dry in summertime. Especially in this part of the island the original top soil has been removed by wind action after the tides have stopped and before the vegetation was established. The wind-blown sands included organic materials originating from the marine benthic organisms, so that notably the the vegetation on the coarser eastern grounds suffered most from nutrient deficiency. It may be significant that just on these grounds succession has passed only two main clusters (2 and 4) contrary to three on all other soils. The western higher grounds have some more clay and, consequently, the vegetation grew better, showing dominance of *Epilobium hirsutum*, *Chamaenerion angustifolium* and grasses such as *Calamagrostis epigejos* and *Festuca rubra*.

Transitions between the two main lines of succession are very few, and occurred only within succession line 2, i.e. plot 10 which passed from cluster 2.1 to 1.2, and plot 35 which crossed from cluster 1.2 to 3.3 and back to cluster 3.1 (see also Table 2).

The succession lines also show characteristics in their sequence of species numbers (Table 4). In series A a first maximum is found occurring in 1970 or later, in series B1 such a maximum arose from 1970 back to 1967, and in series B2 it occurred from 1968 to 1965. The data suggest that on higher grounds, generally drier and poorer in nutrients, peak number of species occur earlier than on lower soils. A second maximum appearing about 1975 is of the same magnitude on an average, but in the permanent plots 20 and 21 (series B2) the number of species is about half. In cluster 1.1 where the environment has changed least after the barrage has been built, the number of species is still increasing very gradually.

The entire data set

The data obtained from 36 permanent plots distributed over the whole area of the Middelplaten have been processed with the same numerical methods as those of which the application is restricted to the Big Island. Table 5 summarizes the results, presenting a table (A), a diagram of cluster transitions (C), and some environmental characteristics (B).

The results indicate that succession started with three vegetation types (clusters 1.1, 1.2 and 1.3) in 1963–64, and diverged to 12 in 1973. These lines of succession can be characterized both floristically and environmentally. There are four main trends in succession.

The first developmental line turns from cluster 1.3 to cluster 2. The plots showing this succession are all situated at the water line of the channels: Mean height above water level in summer is only 3 cm, locally shells are washed ashore (mean coverage 11 %), and soil salinity is rather high (12⁰/₀₀ Cl⁻). The vegetation is composed of halophytes and salt-tolerant glycophytes all the time.

The second line of succession passes from cluster 1.3 to cluster 7.3. It is represented in only one plot (nr. 30), situated about 20 m from the water line in a shallow creek catching rain water. The soil is still brackish as well, but halophytes have nearly all gone owing to long periods of stagnant rain-water, and are replaced by *Alopecurus geniculatus*, *Agrostis stolonifera* and *Eleocharis uniglumis*.

The third line develops through clusters 1.2 and 4.1 to the clusters 6.3, 6.4, 7.1 and 7.4. This line is represented in several plots lying farther from the shore-line and generally situating a little higher above the water level (5–26 cm). Clay content of the soil is of the same order as in the former plots and chlorinity is moderate. Salt-tolerant glycophytes and the most euryhaline halophytes have a great share in these clusters.

The last group of plots develops through clusters 1.1 and 1.3 via the clusters 3 and 5.2 to six final stages which mostly are interconnected indicating that there succession was locally more delayed than in the former successional lines. The plots are situated much higher above the water level of Lake Veere (32–83 cm), and they therefore have a much deeper aerated soil profile. Clay content is much lower and chlorinity dropped soon to minimal values. Local concentrations of shells indicate that the top soil has been blown away in dry periods before it could be colonized by plants. The vegetation is varied but tall herbs are usually dominant or abundant, either well-developed or in stunted forms.

As the three prominent lines of succession are present in all parts of the Middelplaten, vegetational development went in broad outline along the same pathways over the

Table 4. Species numbers in the permanent plots of the Big Island for the period 1963–77.

series	plot	63	64	65	66	67	68	69	70	71	72	73	74	75	76	77
A	9	6	4	4	6	6	6	6	7	6	8	8	9	13	12	13
	34	2	6	7	18	23	28	29	30	31	32	[40]	36	29	24	26
	35	3	13	14	17	18	19	22	[28]	19	15	16	14	16	18	[24]
	33	2	8	13	17	20	22	28	[34	34]	30	23	25	18	[31]	25
	10	16	10	14	22	28	39	36	[40]	36	34	36	36	[45]	41	40
B 1	11	26	28	28	32	30	33	32	37	38	42	[48]	46	42	40	
	17	16	22	20	30	30	34	31	[36]	33	38	44	45	[47]	39	42
	12	17	25	26	34	30	[38	38]	36	33	32	32	34	[39]	32	33
	18	17	25	21	26	29	[37]	24	22	12	13	11	18	[21]	20	19
	19	14	20	19	28	[37]	35	28	22	20	17	17	23	[26]	24	26
B 2	13	14	15	11	15	21	[26]	25	24	23	30	30	35	[38]	33	37
	14	28	22	24	30	33	[38]	35								
	15	18	18	22	26	[29]	28	28			21	16	15			
	21	13	13	19	25	[28]	15	12	9	8	10	8	9	[12]	11	
	20	22	23	[30]	25	20	17	14	13	10	10	11	11	[14]	13	12

Table 5. Table of cluster transitions (A), environmental characteristics (B), and diagram of cluster transitions (C) of 36 permanent plots in the Middelplaten sand flats. P = Peninsula, B = Big Island, S = Small Island.

A   B   C

Area	Plot	Co-ordinates	62	63	64	65	66	67	68	69	70	71	72	73	Numbers of clusters 63-73	Ordnance level (NAP)	Aeration depth (cm)	clay content (%)	coverage (%) of shells	Cl⁻ content (°/oo) 1963	Cl⁻ content (°/oo) 1965	Ordnance level (NAP)	Aeration depth (cm)	clay content (%)	coverage (%) of shells	Cl⁻ content (°/oo) 1963	Cl⁻ content (°/oo) 1965
B	9	248.109		1.3	1.3	1.3	1.3	2	2	2	2	2	2	2	2	0	30	20	35	12	12						
P	27	007.042		1.3	1.3	1.3	1.3	2	2	2	2	2	2	2	2	2	3	28	0	12	12						
P	26	007.055		1.3	1.3	1.3	2	2	2	2	2	2	2	2	2	1	30	18	0	12	12	3	25	21	11	12	12
S	1	397.117		1.1	1.3	1.3	2	2	2	2	2	2	2	2	2	0	25	18	15	12	10						
P	25	011.055		1.3	1.3	1.2	2	2	2	2	2	2	2	2	3	11	35	20	5	10	7						
P	30	033.029		1.3	1.3	1.3	1.3	1.3	7.3	7.3	7.3	7.3	7.3	7.3	2	8	20	25	0	11	11	8	20	25	0	11	11
P	31	032.028		1.2	4.1	4.1	4.1	7.4	7.4	7.4	7.4	7.4	7.4	7.4	3	11	25	24	0	12	11						
P	32	031.028		1.1	1.2	1.2	4.1	4.1	7.4	7.4	7.4	7.4	7.4	7.4	4	14	30	20	10	12	11	13	28	19	5	12	11
S	37	323.178											7.4	7.4			30	12									
P	28	005.042		1.1	1.2	4.1	4.1	4.3	4.3	4.3	7.1	7.1	7.1	7.1	5	10	40	22	0	12	12	18	40	20	1	12	12
P	29	003.045		1.1	1.2	4.1	4.1	4.1	4.2	7.1	7.1	7.1	7.1	7.1	5	26	40	18	2	12	12						
S	8	322.188	1.3	1.1	1.2	3	4.1	4.3	4.3	4.3	6.3	6.3	6.3	6.3	6	13	40	12	5	9	1	13	40	12	5	9	1
B	33	301.107		1.3	1.2	4.1	4.1	4.1	4.1	4.1	6.4	6.4	6.4	6.4	5	5	40	14	0	12	9						
B	34	278.114		1.3	1.2	4.1	4.1	4.1	4.2	4.2	6.4	6.4	6.4	5		20	35	14	0	12	5						
B	35	307.122		1.3	1.2	4.1	4.1	4.1	6.4	6.4	6.4	6.4	6.4	4		30	25	0	12	3	11	33	20	9	12	9
B	10	242.112		1.1	1.2	1.2	4.1	4.1	4.2	4.2	4.2	6.4	6.4	6.4	5	7	25	22	5	12	12						
S	2	398.123	1.3	1.1	1.2	3	4.1	5.2	4.2	4.2	6.4	6.4	6.4	7	11	40	20	40	11	7						
S	36	397.119							6.4	6.4							30	22									
P	23	024.083		1.1	3	3	3	5.1	5.1	5.1	7.2	7.2	7.2	7.2	4	79	60	10	40	1	1	59	53	11	38	2	2
P	24	008.077		1.1	3	3	3	7.2	7.2	7.2	7.2	7.2	7.2	7.2	3	38	45	12	35	3	4						
B	15	214.201		1.1	3	3	3	5.1	5.1	5.1			5.1	5.1	3	60	100	10	10	4	1						
B	14	218.204		1.1	3	3	3	5.1	5.1	5.1					3	56	80	12	30	4	1	69	86	10	28	4	< 1
B	13	238.183		1.1	1.1	1.1	3	5.1	1.1	5.1	5.1	5.1	5.1	5.1	3	76	75	4	70	1	1						
S	6	358.190		1.1	3	3	5.1	5.1	5.1	5.1					3												
S	22	098.086		1.1	3	3	5.1	5.1	5.1	5.1	5.1	5.1	5.1		3	83	90	14	0	8	1	63	63	10	23	2	< 1
S	7	334.181	1.3	1.1	3	3	3	5.1	5.1	5.1	5.1	6.1	6.1	6.1	4	49	60	14	10	3	1						
S	5	378.168	1.3	1.1	1.1	1.1	3	3	5.2	6.1	6.1	6.1	6.1	6.1	4	75	65	5	35	1	1						
S	4	402.143	1.3	1.1	3	3	5.2	5.2	5.2	6.1	6.2	6.2	6.2	6.2	5	46	60	10	35	4	1						
S	3	385.137	1.3	1.1	3	3	5.2	5.2	5.2	6.2	6.2	6.2	6.2	6.2	4	42	50	14	35	4	1	43	51	11	24	6	2
B	12	248.139		1.1	3	3	5.2	5.2	5.2	6.2	6.2	6.2	6.2	6.2	4	52	35	10	35	1	1						
B	11	243.128		1.1	1.1	3	5.2	5.2	5.2	6.2	6.2	6.2	6.2	6.2	5	38	45	15	10	11	3						
B	17	138.129		1.1	1.1	3	4.1	5.2	5.2	5.2	6.2	6.2	6.2	6.2	5	37	45	5	3	12	3	35	45	6	2	12	4
B	18	122.135		1.1	3	3	5.2	5.2	5.2	8.1	8.1	8.1	8.1	6.1	4	32	45	6	0	12	4						
B	19	123.136		1.1	1.1	1.1	5.2	5.2	5.2	8.1	8.1	8.1	8.1	8.1	3	37	45	5	3	12	3						
B	20	127.156		1.1	3	3	8.2	8.2	8.2	8.2	8.2	8.2	8.2	8.2	3	48	60	14	10	11	1	54	65	15	5	11	< 1
B	21	129.155		3	3	3	8.2	8.2	8.2	8.2	8.2	8.2	8.2	2	59	70	15	0	11	1						

| Numbers of transitions | | | | 22 | 8 | 23 | 12 | 8 | 11 | 8 | 3 | 0 | 0 | | | | | | | | | | | | | | |
| Numbers of clusters | | | | 4 | 5 | 5 | 9 | 9 | 11 | 14 | 13 | 12 | 12 | 12 | | | | | | | | | | | | | |

whole area. A more detailed view, however, shows that some vegetation types (clusters) are more confined to one part than to another. Cluster 7, for instance, is nearly restricted to the Peninsula, cluster 6 is found on both islands, and cluster 8 only on the Big Island.

Distribution of single species in space and time

To obtain an idea of the distribution and dispersal of single species diagrams have been computed in which the behaviour of the species in space and time can be read off (Fig. 5). The sequence of the clusters and permanent plots is the same as in Table 5, so that cluster affinity and successional status of the species can easily be defined. Here, only a few species typical for some clusters or groups of clusters, either by presence or by absence, are considered.

Suaeda maritima: a common species in the whole area for the first years after the barrage has been built. *Atriplex hastata* and *A. littoralis* had a similar behaviour. Both *Suaeda* and *A. littoralis* disappeared soon after 1964, also in the brackish environments. *A. hastata* persisted two years longer.

Matricaria inodora: this species developed over nearly the whole area and showed mass development between 1963 and 1966, especially in the more sandy parts (fourth line of succession). After 1970 it decreased rapidly but could persist on many localities (especially the clusters 5.1, 6.1, 6.2 and precursors) in stunted, non-flowering individuals. Species with similar distribution patterns are *Aster tripolium* and *Juncus bufonius* (affinity to the clusters 2, 6.3, 6.4, 7.1–4, and precursors).

Puccinellia fasciculata: a halophyte which is strongly restricted to cluster 2 and its precursor nr. 1.3, and which only occurred temporarily in a few other localities scattered over the area. *Salicornia europaea, Puccinellia maritima, Spergularia media,* showed a similar pattern of distri-

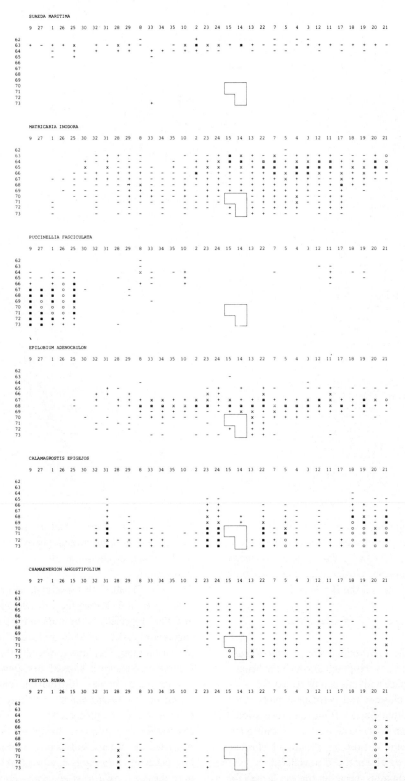

Fig. 5. Distribution over plots in the 1963–73 period of some selected species established in the Middelplaten sand flats. Signatures: white = absent. — = 1–2%, + = 3–5%, x = 6–15%, o $\cong$ 16–45% and $\otimes$ = 46–100% coverage; outlined: no records.

bution. *Puccinellia distans* did so in a less pronounced way.

Epilobium adenocaulon: with the exception of *E. hirsutum* the *Epilobium* species showed mass development between 1964 and 1969. *E. adnatum* one year earlier (1966) than *E. adenocaulon* and *E. parviflorum* (1967–1968). In 1968–69 *Epilobium* seedlings were found in many places. *E. hirsutum* was established in 1966, especially in both islands, and persisted longer than its allies. In the clusters 2 and 7.3, developed in still brackish environments, only very few stunted forms have been found.

Calamagrostis epigejos: this species has dispersed more recently, and persists still, especially in the drier sandy parts. It is characteristic of the cluster 7.2, 8.1 and 8.2.

Chamaenerion angustifolium: strong affinity to the higher sandy soils (fourth succession line) is found in the distribution pattern of *Chamaenerion angustifolium*. Full development, however, is reached in combination with *Hippophae rhamnoides* providing for a nitrogen source.

Festuca rubra: a species with a very distinct pattern of distribution varying from brackish soils to higher sandy soil, and locally dominating. Characteristic of cluster 8.1.

Conclusions

Numerical pattern detection studies in vegetational development on sand flats which became permanently emerged after barrage construction gave the following results:

1. After the barrage had been built desalinization was the deciding environmental process for invasion of plants. Two periods of invasion could be recognized; the first one starting directly after emergence and consisting of halophytes, and, shortly after, the second one consisting of glycophytes. In both pioneer groups annual species predominated. They showed little preference within the range of environmental variation.

2. In the early years the vegetational pattern is mainly temporal (see also Hogeweg 1976). The speed of development of the vegetation differs in different parts of the area, and the observed spatial differentiation is partly due to this difference in speed: e.g. in 1966 much spatial differentiation is partly due to the fact that many different stages are present at the same time (Table 2). The difference in speed of development is, however, correlated with environmental differences (e.g. height above lake level).

3. The order of occurrence of vegetation types in the plots was the same except for divarication of successional lines: neither reversals nor skipping of vegetation types occurred (note that this statement is conditioned by the level of

detailing in classification, though true for optimal subdivisions). However, early succession seems to be regulated so strongly by environmental changes that this fixed order does not seem to indicate the classical type of 'obligatory succession' (see Horn 1976). Previous stages of vegetation do not seem necessary for the later stages, and they are certainly not a sufficient condition for the establishment of later stages (see plot 9, Table 2). Beeftink et al. (1978) found rigid successional sequences in *Halimione portulacoides* clearances. Those rigid sequences of early successional plants seem to be defined by specific biological characteristics determining the strategies for establishment and survival of plants, and by the unidirectional environmental processes induced by disturbance. In the present case these processes include desalinization, nutrient consumption and storage in organic materials, and oxidation of sulphides.

4. In later years the vegetational pattern is mainly spatial (see also Hogeweg 1976) as a result of divarication of the earlier successional lines. This type of succession started 1965–66, and continued to about 1968–69. Hogeweg (1976) concluded that these spatial differences are partly due to 'fossilization' of earlier stages of succession as mentioned above. This phenomenon could be the consequence of an excessive nutrient storage in plant materials, by which the soil is exhausted to below a threshold value essential for establishment of newcomers, and even for maintaining some populations present. Especially perennial herbs, such as *Epilobium* species and *Calamagrostis epigejos*, can play an important role in this respect. Van Andel & Vera (1977) pointed out that in *Chamaenerion angustifolium* high amounts of nutrients are stored and allocated to a large proportion to the overwintering tissues, i.e. the roots. This storage can reduce the availability of mineral nutrients in the soil for other species, such as *Senecio sylvaticus*, so that both *Ch. angustifolium* and *S. sylvaticus* are not capable to re-establish theirselves from seed.

5. Vegetational diversity expressed in numbers of clusters reaches a maximum in 1969–70, i.e. 9–10 years after the barrage had been built. Contrary to the Big Island subset the whole data set shows a tendency to a slight decrease in diversity in the following years. This suggests that the first vegetational development has ended, and further successional trends have to be initiated by other, long-term ecological processes in which the vegetation will participate. These may be humification, decalcification, and acidification processes in the topsoil layers of which the

85

first tendencies have already been recognized (Koutstaal & Sipman 1977).

6. The total number of species increases during the period of observation (Hogeweg 1976) but in the single plots peak densities in species numbers are found alternating with periods with lower numbers of species. The first peak occurred 4–10 years after the barrage had been constructed. The lower and moister the soil is, the later the peak arose, and the salt vegetation at the islands' margins still did not show any peak in species number till now. These differences in the temporal pattern are probably related to differences in soil processes started up by the barrage construction. The second peak density seems to be a more concurrent one, and occurs 5–10 years after the first one. It is less evident on the higher, sandy grounds where the vegetation impoverished more than on the lower ones, and recovery by species settling took more time. The occurrence of a sequence in peak numbers of species may be caused by the impact of environmental disturbance induced by construction of the barrage. A similar effect has been observed by Watt (1947) and Williams et al. (1969) after deforestation. Such an impact will initiate – it can here only be hypothesized – simultaneity in establishment and break down in plant life.

7. Basically, successional trends are similar in all parts of the area of investigation. Because of the size of the area involved differences in floristic composition and structure between three parts should be the result of environmental differences rather than of differences in accessibility for invasion of plants. Although succession and spatial variation in the three parts of the area of investigation were similar, all plots of the Peninsula (including both wet and dry plots) could be distinguished from all Island plots from 1968 onward, using the technique of iterative character weighing (Hogeweg 1976). This demonstrates that similar successional trends may occur despite of consistent differences in vegetation and environment (including grazing). This result indicates again that succession can only be weakly dependent on previous stages, and may perhaps even be a relatively independent process all together.

Summary

Vegetation succession on the tidal flats of the Middelplaten, isolated from tidal action after the construction of a barrage in 1961, has been studied with cluster analysis. Data of yearly cover-abundance estimates of species over 1963-73 produced from the Big Island have been partitioned at two levels yielding 7 and 17 clusters respectively. Mapping the 17 clusters for the successive years showed pathways in cluster distribution related with the desalinization rate and water relations in the soil. From the cluster transitions three successional lines could be recognized. They represent developments under different environmental conditions in which differences in desalinization rate, water relations and availability of nutrients are deciding.

In the whole area of investigation two invasions could be recognized in the period of colonization, one of halophytes, and the other of glycophytes. In the early years the vegetational pattern is mainly temporal, probably related with unidirectional environmental processes and strategies of species. In later years the pattern is mainly spatial, partly due to 'fossilization' of earlier stages of succession for which reduction in availability of mineral nutrients may be responsible.

Vegetational diversity expressed in numbers of clusters reaches a maximum about 10 years after the barrage had been constructed. It is suggested that a second period of succession will be initiated by long-term ecological processes, such as humification, decalcification and acidification, in which the vegetation will participate.

References

Andel, J. van & F. Vera. 1977. Reproductive allocation in Senecio sylvaticus and Chamaenerion angustifolium in relation to mineral nutrition. J. Ecol. 65: 747–758.

Beeftink, W.G., M.C. Daane & W. de Munck. 1971. Tien jaar botanisch oecologische verkenningen langs het Veerse Meer. Natuur Landschap 25: 50–63.

Beeftink, W.G., M.C. Daane, W. de Munck & J. Nieuwenhuize. 1978. Aspects of population dynamics in Halimione portulacoides communities. Vegetatio 36: 31–43.

Doing Kraft. H. 1954. L'Analyse des carrés permanents. Acta Bot. Neerl. 3: 421–424.

Hogeweg, P. 1976. Ecological application of pattern detection methods, in: P. Hogeweg, Topics in biological pattern analyses, p. 151–199. Thesis Utrecht.

Hogeweg, P. & B. Hesper. 1972. BIOPAT, program system for biological pattern analysis. Bioinformatica Group, University of Utrecht.

Horn, H.S. 1976. Succession, in: May R.M. (ed.) Theoretical ecology, p. 187–204. Blackwell, Oxford.

Kendall, M.G. 1972. Cluster analysis, in: Watanabe, S. (ed.) Frontiers of pattern recognition, p. 291–310. Academic Press, New York.

Koutstaal, B.P. & H.J.M. Sipman. 1977. De korstmossen van de Middelplaten. Levende Natuur 80: 248–260.

Noordwijk-Puijk, K. van. 1976. De vegetatie op de Middelplaten in 1973 en een analyse van het ontwikkelingsproces. Delta Institute for Hydrobiological Research, Yerseke, Student Reports D3-1976, 215 pp.

Ward, H.H. 1963. Hierarchical grouping to optimise an objective function. J. Amer. Stat. Ass. 58: 236–244.

Watt, A.S. 1947. Pattern and process in the plant community. J. Ecol. 35: 1–22.

Williams, W.T., G.N. Lance, L.J. Webb, J.G. Tracey, & M.B. Dale, 1969. Studies in the numerical analysis of complex rain-forest communities. III. The analysis of successional data. J. Ecol. 57: 515–535.

Wishart, D. 1969. An algorithm for hierarchical classifications. Biometrics 22: 165–170.

MIGRATION AND COLONIZATION BY VASCULAR PLANTS IN A NEW POLDER*,**

W. JOENJE***

Rijksuniversiteit Groningen, Biological Centre, Department of Plant Ecology, Haren (Gr.), The Netherlands

Keywords:
Colonization, Desalination, Dispersal, Halophyte, Life form, Sandflat

Introduction

The migration of species into a newly embanked bare area concerns the pre-stage of succession, during which great emphasis lies on the accessibility of the area and the dispersal types of the plants. Other characteristics of the life-cycle, together with environmental factors, determine the success of an ecesis (establishment of a species on an unoccupied site). It often remains unknown to what extent the ecesis is hampered by adverse environmental conditions. The arrival of diaspores can only seldom be ascertained directly. Perhaps the best information can be expected from new, mesic environments, suitable for the ecesis of a great number of species.

As to the migration of plants into new polders, Feekes & Bakker (1954) discerned three stages, with respect to the dispersal of propagules:

1. the arrival of propagules before the drainage, mainly by water,

2. the arrival on the newly exposed land, mainly by wind and water,

3. the dispersal caused by human activities, related to reclamation and exploitation.

In several studies the predominant role of anemochorous dispersal is stressed (Ulbrich 1928, Salisbury 1954). Bakker & van der Zweep (1957) compared dispersal spectra from 10 year old vegetations near Urk in the North-Eastern Polder to those of the nearest old vegetations, and demonstrated the major role of eu-anemochorous species (*sensu* Westhoff 1947). These authors concluded that at least in western Europe and in non-saline environments the eu-anemochorous dispersal (type A, Table 1) is pre-eminently suited for young and still open vegetation.

The migration and subsequent colonization of a polder under desalinating conditions renders a different picture, mainly because only a few halotolerant eu-anemochorous species persist here. During 5 years in the saline Wieringermeerpolder, the role of eu-anemochorous species, although important, remained less than was expected (Feekes 1936). This was probably caused by a failing ecesis in the saline environment, although desalination took place rather quickly, caused by drainage channels dredged before the enclosure. Also the relative success of hydatochorous dispersal is mentioned.

In the recently embanked Lauwerszeepolder (Netherlands) no drainage systems were constructed and the desalination proceeded at a slower rate. For 6 years the development of spontaneous vegetations on the drained sandflats was investigated. For a general description and quantitative data on vegetation and soil see Joenje (1974). This paper summarizes floristic data and discusses migration, ecesis in the first, second and sixth year, in three habitat types of the flats: musselbanks, sand and silty

* Nomenclature of *Salicornia* spp. follows Tutin et al., 1964, Flora Europaea, Vol. I; other plant taxa Heukels & van Ooststroom 1975, Flora van Nederland, 18e druk, Wolters-Noordhoff, Groningen.
** Contribution to the Symposium of the Working Group for Succession research on permanent plots of the International Society for Vegetation Science, held at Yerseke, The Netherlands, October 1975.
*** The author wishes to thank Mr J.D.D. Hofman and Mr J. Franke for their share in the fieldwork and preparation of the species inventory, Prof. Dr D. Bakker for his interest and criticism, Mr E. Leeuwinga for preparing the figures, and Mrs A. Severijnse for improving the English. The author is also grateful to the Rijksdienst voor de IJsselmeerpolders, who kindly provided facilities, and to the Rijkswaterstaat for supplying transport. The work was part of a study, sustained by the Netherlands Organization for the Advancement of Pure Research (Z.W.O.).

sand. Changes in the life-form spectra are considered. Furthermore, the resulting colonization stages are described and illustrated with data of three permanent plots over 6 and 7 years.

Environmental characteristics of the Lauwerszeepolder

In May 1969 the Lauwerszeepolder (Fig. 1), a wide estuary of the Dutch Waddensea, was cut off from tidal influence. From that time the tidal channel system has formed a water reservoir of about 2000 ha receiving the surplus precipitation of the catchment area (40,000 ha) of three rivers and three other discharges. The emerging surface of 6100 ha mainly consists of extended sandflats (4500 ha) divided by the channels and bordered by a furrow-system of land reclamation works along the former coast. These areas now serve agricultural purposes, while the sandflats remained undisturbed and gained a nature reserve status. Due to level fluctuations in the reservoir, a large area of the flats is occasionally flooded during winter.

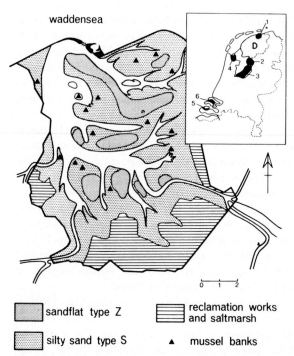

waddensea

| | sandflat type Z | | reclamation works and saltmarsh |
| | silty sand type S | ▲ | mussel banks |

Fig. 1. Map of the Lauwerszeepolder showing, a.o., the position of the tidal channels, silty sands, sands and musselbanks.
Inset: The Lauwerszeepolder and the discharge area (D) and some other embankments in the Netherlands. 1. Lauwerszeepolder; 2. North-Eastern Polder; 3. Flevopolders; 4. Wieringermeerpolder; 5. Veerse Meer; 6. Grevelingen Meer.

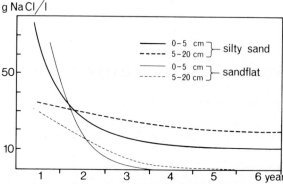

Fig. 2. Desalination in the upper soil layer in the course of six years on sandflats and silty sand in the Lauwerszeepolder.

The flat sands consist mainly of fine grained sediments (90–140 μ) with only small amounts of silt $< 16 \mu$ (2–5%) and 3–6% carbonate; the amount of organic matter is $< 1\%$. The fine texture and the lack of local drainage caused a slow percolation of precipitation resulting in an unexpectedly slow desalination (Veenstra 1973); the soil was often saturated, especially in winter, and surplus water ran off superficially; probably most of the salt was removed by diffusion (Joenje, unpublished data).

During the first years after the embankment the salt concentration in the upper soil layer rose sharply to high values in dry periods (e.g. 50–150 g NaCl/liter soil moisture).

The surface layer of the coarser, practically siltless parts of the sandflats was blown away in some places, offering quite another extreme. These coarser sands, however, desalinated much quicker (Fig. 2), especially when situated higher and closer to the channels, thus presenting glycophyte habitats from the second year on (highest NaCl concentration below 5 g/l).

A third quite deviating habitat type was found on the musselbanks, scattered accumulations of mussels (*Mytilus edulis*) of 1 to 30 ha, in a silty deposit, with a height of 1 to 15 dm. Here desalination proceeded very quickly, permitting glycophytes to establish already in the first year. Because of their rough surface (catching propagules), relief and dark colour (providing shelter and higher temperatures), high clay content, and a high soil fertility caused by the decaying marine fauna, these structures probably offered a very attractive, mesic environment. For these reasons and because of the long distance (mostly $\gg$ 1 km) to the nearest vegetation, the musselbanks can be regarded as excellent 'species traps' suitable for migration studies.

Fig. 1 gives the occurrence of the habitat types described.

90

Origin of propagules and migration vectors

Arrival of propagules from widely different habitats could be expected. Anemochorous dispersal was probable from the former salt marshes, the surrounding mainland (E and SE of the polder mainly arable land, W and SW pastures) and from the nearby dune islands (Schiermonnikoog at 8 km); hydatochorous arrival was possible via the discharge to the reservoir, draining a vast area (Fig. 1). Drift material, especially floating stolons and sods of *Phragmites australis* and *Scirpus maritimus*, together with smaller species of *Phragmition* and *Bidention* communities were found along the bare edges of the new lake. Moreover, species from the Wadden area, washed ashore in drift material at the seaward side of the enclosure dam, could be blown into the polder. Exo- and endochorous dispersal by birds should be mentioned as a possibility. Many species of waders, ducks (esp. *Anas penelope* and *Anas cracca*) and geese (*Branta leucopsis*) frequented the area in winter; the polder played an important role in their migratory movements. During the first years, avocets, terns and gulls, especially the black-headed gull, established breeding colonies on the musselbanks.

The existence of resident seed populations was demonstrated in some samples from the topsoil layer, taken prior to the embankment. The result points to a predominantly hydatochorous dispersal.

Plant migration

First it must be stated that a discussion of euchorous dispersal is especially relevant when restricted to the first and second year. Later on, the data of different habitats become less informative because of increasing short-distance dispersal from previously colonized areas. In 1969, actual dispersal was occasionally observed, e. g. the pappus of *Senecio congestus* on the musselbanks, seeds of *Betula* sp. in window traps for insects (J. Meyer, pers. comm.) and vegetative parts of several species along the edges of the reservoir and higher, deposited by water transport.

The interesting question as to what extent the different dispersal types reach the polder early will be investigated on the basis of the data of repeated floristic inventarisation of the three habitat types. In Table 1 the 71 important phanerogams are given, together with their relative abundances, local dissemination types and life forms. In the actual dissemination of many euchorous species, often more than one way of transport is involved. To avoid many speculations connected with the use of a more detailed classification, we reduced the five euchorous types mentioned by Westhoff (1947) to two types 'A' and 'H' and added a local type 'O', meaning a combination of dispersal by wind and water. The latter type played a very important role in the flat and bare environment, where material was transported by wind or by rainpools and lake water being blown over. A fourth group 'R' comprises the eu-zoochorous (anthropochorous) and the remaining dyschorous species. The characterization of the species is mainly based on field evidence; some use has been made of the classification and critical discussions of Feekes (1936). Nevertheless the classification raised difficulties: the 'O' type could well have been underestimated in favor of type 'A', (e.g. *Aster tripolium*, *Phragmites australis*); according to Feekes, *Cirsium arvense* was listed as A-type, although there is evidence of a R-type dispersal.

The dispersal spectra derived from the complete species list comprising 227 species from the three habitats for the first, second and sixth years are given in Table 2, showing the number of species. Fig. 3 gives the spectra of musselbanks and sandflats, based on percentages. *

It appeared that on the mesic musselbank habitat species could establish in the first two years. The dominance of hydatochorous (H-) species is most probably caused by dispersal prior to the drainage. The seeds of many species, transported by the descending tidal and the discharged fresh waters, could well have reached the musselbanks. Most of the species found can retain their viability in seawater for many months to more than a year (Feekes 1936).

In the first year, a group of annual halophytes or halo-tolerant species of the lower saltmarsh and tidal drift deposits was dominating (viz. Table 1), but also several species from the mainland as *Rumex crispus*, *Polygonum lapathifolium*, *Bidens tripartitus* were found. All these species are common on the surrounding mainland. Most of the anemochorous (A-) species only settled at the end of the first growing season. It is, however, probable that they had partly arrived before the enclosure.

In the second year, the musselbanks offered a rather bare and still better aerated and fresh environment. While direct hydatochorous transport must be excluded now, the observed increase from 15 to 33 H-species is striking.

* The complete list is available on request from the Department of Plant Ecology, Haren.

Table 1. List of the most important Phanerogams found in three habitats in the Lauwerszeepolder in the first, second and sixth year. Columns: L life form, D dispersal type, M musselbanks, Z sandflats, S silty sands; life form after Raunkiær (T therophytes, i.e. ephemerals summer and winter annuals, He hemicryptophytes, C chamaephytes, G geophytes, P phanerophytes); local dissemination types (A (eu-anemochorous) propagules dispersed by wind over some hundred metres to many kilometres, H (eu-hydatochorous) propagules dispersed by water, floating for some hours to years, O (combination of A and H) typical for the bare and flat environment where rainpools and drift material are blown across the surface, R (rest group, consisting of eu-zoochorous, anthropochorous and dis-chorous species); abundance (+ one or a few individuals, (r)are, (o)ccasional, (f)requent, (a)bundant, (d)ominant).

Year	L	D	1969 M	1969 Z	1969 S	1970 M	1970 Z	1970 S	1974 M	1974 Z	1974 S
Salicornia dolichostachya	T	H	+						o		
S. europaea	T	H	+			o	o	f	d	d	o
Atriplex hastata	T	H	+	+	r	o	o	o	d	d	
Suaeda maritima	T	H	+			o	o	f	d	d	
Puccinellia maritima	He	H	+	+	+	f	o	f	a	a	a
Atriplex litoralis	T	H	+				r		f	a	o
Ranunculus sceleratus	T	H	+	+		r	o	o	r	r	
Matricaria maritima	T	H	+	+	+		o		r	r	
Polygonum l. ssp. inodora	T	H	+								
P. l. ssp. lapathifolium	T	H	+				r				
Aster tripolium	He	A	+			r	+		f	a	f
Phragmites australis	G	A	+	+	+	o	r		f	o	o
Senecio congestus	T	A	+			f	f	f	o	f	
S. vulgaris	He	A	+				o	+	o	o	
Taraxacum sp.	T	A	+			o					
Sonchus asper	T	A	+			f	f	f	o	o	
S. oleraceus	T	A	+								
S. arvensis	G	A	+								
Cirsium vulgare	G(He)	A	+			o	o		r	o	
C. arvense	G	A	+			o	o		d	o	o
Chamaenerion angustifolium	G	A	+			f	f	f	f	o	
Epilobium hirsutum	T(He)	A	+		+	f	f		d	f	f
Poa annua	He	O	+			o	o		a	a	a
P. trivialis	He	O	+			o			a	a	a
Lolium perenne	He	O	+			o	o		a	a	a
Stellaria media	T	O	+			o	o	r	o	o	o
Spergularia media	He	O				f	f		f	f	f
Rumex crispus	He	O	+			o	o		o	o	
Alopecurus geniculatus	He	O	+			r			r	r	
Agrostis stolonifera	He	O	+			r	r		f	f	r
Cardamine pratensis	He	O	+			r			r	r	o
Mentha aquatica	He	O	+			o			o	o	
Polygonum persicaria	T	O	+			o	o		o		
P. lapathifolium ssp. pallidum	T	O	+					r	o		+
Epilobium parviflorum	G	A	+					r	o	o	

Year	L	D	1969 M	1969 Z	1969 S	1970 M	1970 Z	1970 S	1974 M	1974 Z	1974 S
E. adnatum ssp. adnatum	G	A							o	r	
Erigeron canadensis	T	A								r	
Tussilago farfara	G	A							o	o	
Leontodon autumnalis	H	A						r		r	r
Typha latifolia	G	A									
Salix alba	P	A						r			
Puccinellia distans	H	O				+	+	+	+	+	
P. capillaris	H	O				+	+	+		+	
Calamagrostis epigejos	H	O				+	+	+		+	
Elytrigia repens	H	O				+	+	+		+	
Festuca arundinacea	H	O				+	+	+	o	+	
F. rubra	H	O				+	+	+		+	a
Holcus lanatus	H	O				+	+	+		o	
Juncus bufonius	H	O							+	+	+
J. gerardi	G	O							+	+	+
Cerastium holosteoides	T(H)	O							+	+	+
Sagina maritima	T	O							+		
Spergularia marina	T	O						+		o	a
Odontites verna	T	O							+	+	
O. verna ssp. serotina	T	O								r	r
Rumex acetosa	He	O									
Urtica dioica	He	R						o		o	
Hippophae rhamnoides	P	R									
Polygonum aviculare	T	R				+	+		r	r	
Gnaphalium uliginosum	He	R				+	+			r	
Plantago maritima	He	R				+	+	+	+	+	r
Triglochin maritima	He	R				+	+		o	+	+
Salix repens	P(C)	R									
Angelica silvestris	He	R								r	
Lycopus europaeus	He	R							o	o	r
Valeriana officinalis	He	O							r	r	
Juncus articulatus	He	O							f	f	r
Sagina nodosa	He	O								r	o
Calamagrostis canescens	He	R							+	+	
Carex acutiformis	He	R							+	+	+
C. otrubae	He	R							+	+	+
Lythrum salicaria	He	R							+	+	

Table 2. Local dispersal spectra of the flora of three habitats in the first, second and sixth year after enclosure (for explanation of dispersal types and habitat viz. table 1).

Year	1969			1970						1974					
	total			new in polder			total			new in polder			total		
Habitat	M	Z	S	M	Z	S	M	Z	S	M	Z	S	M	Z	S
Dispersal type															
A	12	1		20	2		32	6		3	6		26	30	1
H	15	6	6	33	4		58	15	8	5			22	25	14
O	8			9	3		17	7	1	3	4		17	17	1
R	2			68			69			12	18	1	53	54	7
Number of species		39			178						177				
New species		39			141						48				

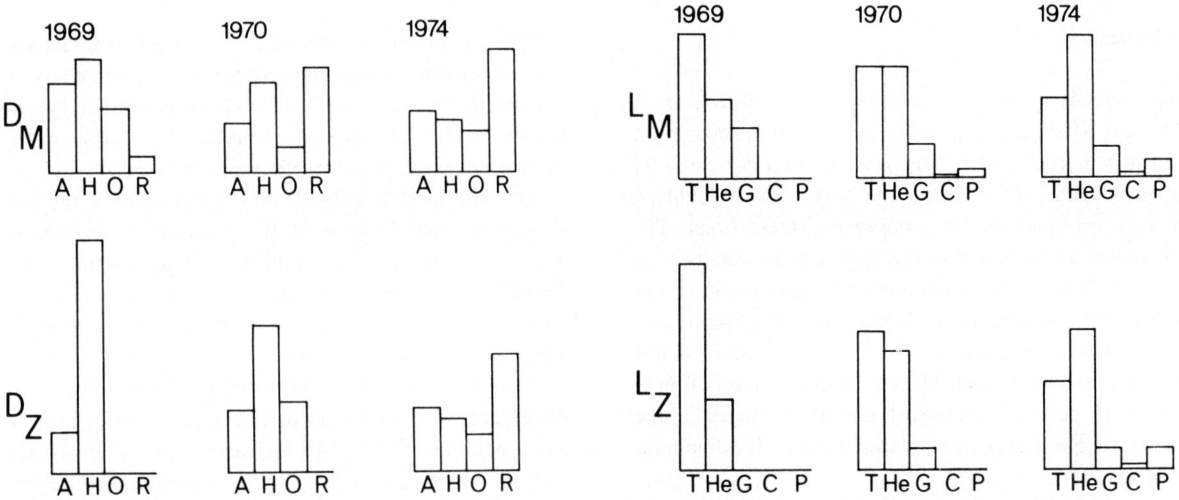

Fig. 3. Dispersal spectra of musselbanks (D_M) and sandflats (D_Z), and life form spectra of musselbanks (L_M) and sandflats (L_Z), in percentage of species.

The ecesis from a resident seed population or the arrival by means of a combined wind-water dispersal (type 'O' could explain this, meaning an underestimation of the last type. The origin of many species listed under type 'R' is equally questionable; among these, many species with hard-scaled fruits and seeds were present.

After six years the total number of species on the musselbanks had decreased considerably. The vigorous growth of species as *Epilobium hirsutum*, *Chamaenerion angustifolium*, *Cirsum vulgare*, *C. arvense* and *Poa trivialis* and *Agrostis stolonifera* prevented many smaller species to persist, especially therophytes. The open and sparse vegetation of the more or less desalinated sands (Z) now consists of even more species. By now the spectra of these two habitats show much resemblance. On the saline sands (S) only halotolerant species thrived, mostly of the H-type, reflecting a continuation of the Wadden-environment. This dispersal type may be regarded as characteristic for the halosere of the Waddensea-area (Westhoff 1947) and the species list comprises much of the flora of sandy salt marshes and beach plains of the dune islands (e.g. Joenje & Thalen 1968).

93

Table 3. Life form spectra of the flora of three habitats in the first, second and sixth year after enclosure (for explanation of life form type and habitat viz. table 1).

Year	1969				1970				1974			
Number of species	39				178				177			
Habitat	M	Z	S	total	M	Z	S	total	M	Z	S	total
Life form												
T	21	6	4	21	74	13	6	75	37	42	7	56
He	11	2	2	11	73	11	4	75	67	67	16	88
G	7			7	23	2		23	15	10		16
C					2			2	1	2		3
P					6			5	8	10		14

Colonization

The stages of migration and ecesis were followed by a colonization stage during which growth and propagation lead to a marked species aggregation. Limited by habitat features, each species sooner or later became involved in intra- or inter-specific competitive interactions. The vegetation succession on three permanent quadrats of 5 × 5 m2 offers a clear illustration of this process on slower and faster desalinating areas (Table 4). Quadrat I represents habitat type S, quadrat III lies in habitat type Z and quadrat II gives a transitional habitat type. Based on data of these and other quadrats, the most important stages in the colonization process may be characterized by the dominant species.

Musselbanks: annual stage with *Atriplex hastata, Senecio vulgaris, Polygonum lapathifolium* – perennial stage with *Cirsium vulgare, C. arvense, Epilobium hirsutum, Chamaenerion angustifolium,* several grass species gaining dominance.

Sandflats: only in the second year: annual stage with *Salicornia dolichostachya, S. europaea, Atriplex hastata* – *Puccinellia capillaris, Poa annua, P. trivalis,* – *Epilobium* sp. div., bryophytes* and many species of grasses and forbs.

Silty sands: annual stage with *Salicornia europaea, S. dolichostachya* and *Suaeda maritima* – *Spergularia marina* – *Spergularia media, Puccinellia maritima.*

* The interesting colonization by the euchorous group of bryophyte species is discussed in a separate article (Joenje & During 1977).

The stages in early succession show much resemblance to those of other saline embankments along the coast of the North Sea, e.g. the Hauke-Haien-Koog (Brehm & Eggers 1974), the Meldorfer Bucht, the Veerse Meer (Beeftink et al. 1971) and the Grevelingen Meer.

Life-form spectra (Table 3 and Fig. 3) may offer another ecological generalization of the colonization processes. The classification after Raunkiaer clearly demonstrates the shift from therophytes towards a dominance of hemicryptophytes. This phenomenon, well-known from early stages of succession, may demonstrate the fugitive or r-selected qualities of early-stage therophytes (viz., e.g., old field successions, Mellinger & McNaughton 1975, Tramer 1975, Schmidt 1975). The increasing spread of species over all five classes reflects an ecological differentiation in both the habitats S and M. However, in the sixth year the musselbanks showed a marked reduction in the number of species, especially therophytes and geophytes. Although the two former habitats are quite different, after six years the life-form spectra had become much alike. The increase in hemicrytophytes and other life forms is another aspect of the ecosystem development, which is further accompanied by an increasing repartition of growth and flowering periods over the year, also reflected in the list of species (Table 4). These phenomena, however, can only be discussed on the basis of local information on soil and species. Furthermore, the ecological significance of both types of spectra should increase when related to the extent (frequency, biomass, of the contributing species and then, a further differentiation of the dispersal and life form types would be useful.

Table 4. Phanerogam and Bryophyte species on permanent quadrats (5 × 5 m^2) in the Lauwerszeepolder, found in the first 6 or 7 years after enclosure. Estimation of frequency in 16th. Environmental description in text.

			I						II					III		
Quadrat number			I						II					III		
Coverage % phanerogams	12	20	15	25	25	20	20	20	25	25	55	85	95	90
Coverage % bryophytes												+	20	10
Year after drainage	3	4	5	6	7	3	4	5	6	7	3	4	5	6
Salicornia dolichostachya	7	15	1			8	16	12						
S. europaea	16	16	16	16	16	16	16	16	16	16	16	16	12	
Atriplex hastata	3	12	8	4	1	5	15	10	12	6	16	16	14	6
Suaeda maritima	15	16	16	4	8	5	14	16	5	8	8	10	5	1
Aster tripolium	3	10	16	16	16	1	10	12	16	16	9	16	16	12
Puccinellia capillaris		1	1	5	1						1	2		
Spergularia marina		14	16	16	16		13	16	12	16		9	10	5
Puccinellia distans		2	7	13	15		1	6	10	13		1	8	5
P. maritima				1	1	4	5	6	14					
Juncus bufonius				1		1	3	2	16					
Poa annua		1						2	5	2		7	16	6
Rumex crispus				1					1	4				2
Stellaria media								1						3
Matricaria inodora				1				2	2	5			5	
Poa trivialis								1	1	9		10	16	16
Juncus articulatus										1				
Agrostis stolonifera				3						8		1	13	8
Taraxacum sp.		1										9	10	14
Odontites verna serotina				1						2				
Ranunculus sceleratus												15		
Sonchus arvensis												12	16	2
Senecio vulgaris											14	8	14	4
Epilobium hirsutum												15	16	15
E. adnatum												8	15	8
E. parviflorum												8	16	9
E. obscurum													8	
Cirsium arvense													14	8
Cerastium holosteoides													2	6
Rorippa islandica												1	1	
Phragmites australis												2	7	12
Poa pratensis												1		
Spergularia media													3	
Mentha aquatica													1	3
Angelica silvestris														1
Carex distans														2
Lycopus europaeus														1
Festuca rubra														3
Funaria hygrometrica											16	15	4	
Ceratodon purpureus												5	8	8
Bryum bicolor												8	10	8

Summary

The migration and subsequent colonization in three habitats of the recently reclaimed Lauwerszeepolder, the Netherlands, is described. Origin of propagules, accessibility and other environmental features determining the success of ecesis, are briefly discussed. Dispersal spectra and life form spectra of the flora of three habitats in the first, second and sixth years are compared.

Regarding migration, the data can be summarized as follows:
- Eu-hydatochorous dispersal played a dominant role in all habitats described.
- Probably the seed populations of many species were present at the time of the enclosure.
- After six years anemochorous species dominated on the musselbanks and the desalinated sandflats, not only because of their increase, but also because many hydatochorous species disappeared.
- The dispersal spectra of the remarkably different habitats M and Z showed a great resemblance after six years.
- On the musselbanks, the largest number of species was found in the second year.
- A still larger number of species was found on the desalinated sands after six years.

The colonization and the first successional trends can be summarized as follows:
- In the first year, therophytes dominated.
- A marked shift from therophytes towards hemicryptophytes was found in all habitats. Only a small number of species belonged to other life forms.
- The life form spectra of the remarkably different habitats M and Z showed a great resemblance after six years.

References

Bakker, D. & W. van der Zweep. 1959. Plant migration studies near the former island of Urk in the Netherlands. Acta Bot. Neerl. 6: 60–73.

Beeftink, W.G., M.C. Daane & W. de Munck. 1971. Tien jaar botanisch-oecologische verkenningen langs het Veerse Meer. Natuur en Landschap 25: 50–63.

Brehm, K. & T. Eggers. 1974. Die Entwicklung der Vegetation in den Speicherbecken des Hauke-Haien-Kooges (Nordfriesland) von 1959 bis 1974. Schr. Naturw. Ver. Schlesw.-Holst. 44: 37–36.

Feekes, W. 1936. De ontwikkeling van de natuurlijke vegetatie in de Wieringermeer-polder, de eerste droogmakerij van de Zuiderzee. Diss. Amsterdam. Ned. Kruidk. Arch. 46, 295 pp.

Feekes, W. & D. Bakker. 1954. De ontwikkeling van de natuurlijke vegetatie in de Noordoostpolder. Van Zee tot Land 6, 92 pp.

Joenje, W. 1974. Production and structure in the early stages of vegetation development in the Lauwerszeepolder. Vegetatio 29: 101–108.

Joenje, W. & H.J. During. 1977. Colonisation of a desalinating Wadden-polder by bryophytes. Vegetatio 35: 177–185.

Joenje, W. & D.C.P. Thalen. 1968. Het Groene Strand van Schiermonnikoog. De Levende Natuur 71: 97–107.

Mellinger, M.V. & S.J. McNaughton. 1975. Structure and function of successional vascular plant communities in central New York. Ecol. Monogr. 45: 161–182.

Salisbury, E. 1954. Weed dispersal and persistence. Proc. Br. Weed Control Conf. 1, pp. 289–295.

Schmidt, W. 1975. Vegetationsentwicklung auf Brachland – Ergebnisse eines fünfjährigen Sukzessions-Versuches. In: W. Schmidt (ed.). Sukzessionsforschung. Ber. Int. Symp. 1973 der Int. Ver. für Vegetationskunde, pp. 407–434. Cramer, Vaduz.

Tramer, E.J. 1975. The regulation of plant species diversity on an early successional old-field. Ecology 56: 905–914.

Ulbrich, E. 1928. Biologie der Früchte und Samen (Karpobiologie). Biol. Studienbücher, Julius Springer Verlag, Berlin.

Veenstra, K. 1973. De ontzilting van de gronden in de Lauwerszee in 1971 en 1972. Internal. Comm. Rijksdienst voor de IJsselmeerpolders No. 315, Baflo, 31 pp.

Westhoff, V. 1947. The vegetation of dunes and salt marshes on the Dutch islands of Terschelling, Vlieland and Texel. Thesis, State Univ. Utrecht, 131 pp.

CHANGES IN A SALT-MARSH VEGETATION AS A RESULT OF GRAZING AND MOWING – A FIVE-YEAR STUDY OF PERMANENT PLOTS*, **

J.P. BAKKER***

State University Groningen, Biological Centre, Department of Plant Ecology, P.O. Box 14, Haren (Gr.), the Netherlands

Keywords: Grazing, Management experiment, Mowing, Salt-marsh, Vegetation dynamics

The area

Schiermonnikoog is one of the Dutch Frisian Islands, located in the northeastern part of the Netherlands (53°30' N.L., 6°10'E.L.). It consists of a body of older sanddunes in the northwest and an embanked former salt-marsh (Banckspolder) area south of it. Eastwards of this area, a vast, originally sandy plain extends in the northern part of which small sanddunes developed, while in the south a salt-marsh (Oosterkwelder) has been formed (Fig. 1). Geomorphological details can be found in Bleuten (1971). The work to be discussed was carried out in the Oosterkwelder area, a relatively young salt-marsh. The landscape as a whole is a very dynamic continuously changing one. For the vegetation this can easily be confirmed by comparing data in older references (den Hartog 1952, Westhoff 1954) with the present situation.

History of management

For many years before 1958, young cattle owned by farmers from both the island and the mainland grazed daily in the

dunes and on the salt-marsh during the summer half year. In 1958, a fence was constructed limiting grazing to a particular part of the salt-marsh. From 1958 to 1971 the fenced area included 51 ha situated in the western part of the salt-marsh, combined with a cultivated pasture of 9 ha inside the dike of the Banckspolder where fresh drinking water is available. About 20 of the 51 ha is fertilized twice a year. In the opinion of the authority in charge of the management ('Dienst der Domeinen', a Service of the Finance Department) the vegetation of the remaining areas of the salt-marsh would become rugged when not grazed. At the request of the farmers on the island for extending the grazing area by 32 ha in 1970, the authority readily agreed. From 1972 onwards, between 120–150 young cattle have grazed this area of ca. 90 ha. This means a grazing pressure of 1.3–1.7 animals/ha for the period from the end of May until September/October. This pressure is relatively high and implicates a short-term production of domestic animals rather than conservation purposes (Beeftink 1977).

Framework and objectives of the study

Against the above background, the Department of Plant Ecology started a study-programme in 1971 to follow the changes in the species composition of the vegetation as a result of the renewed grazing policy. The study fits in a national grazing research scheme of the Research Institute for Nature Management (Oosterveld 1975).

In 1971 the original situation of the area of 32 ha, now again under grazing, was laid down. In 1972 a fence was placed enabling a comparison of the vegetation response to the different managing practices inside and outside. Parts of the non-grazed area of the salt-marsh are more or less regularly mown by farmers and the authority in charge of

* Nomenclature of taxa follows Heukels & van Ooststroom (1970), that of syntaxa Westhoff & den Held (1969).
** Contribution to the Symposium of the Working Group for Succession research on permanent plots of the International Society for Vegetation Science, held at Yerseke, the Netherlands, October 1975.
*** Thanks are due to Prof. Dr. D. Bakker for his interest in the present study and his critical reading of the text, and to Drs D.C.P. Thalen for critical reading and correction of the English text. Thanks are also due to the students who analyzed the permanent plots: M. van der Duim and F. Prins (1971), M. van der Duim (1972), R. Schwab and T. Schwab-Vos (1973), and G. Allersma (1974, 1975).

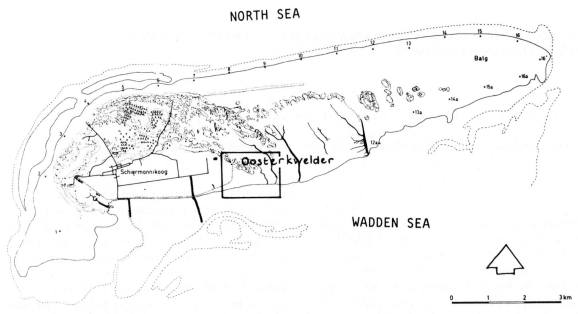

Fig. 1. Map of Schiermonnikoog, with location of the Oosterkwelder.

management. A number of mowing experiments were therefore included in the study in order to place the effect of grazing in a broader framework, namely comparing it with other management practices.

The main objective of this study was to follow and possibly explain the effect of different management regimes viz. grazing, mowing (at different dates) and the combination of the two, on different types of salt-marsh vegetation. This objective includes both a fundamental ecological and a practical aspect. Such data should, in the long run, form a base for improved vegetation management, enabling proper action once the management objectives are defined.

Methods

In order to simulate the different management regimes, a number of mowing treatments (including not mowing) have been carried out in plots of 5 × 8 m since 1971, on both sides of the fence (see Tables 2–6). Areas which were considered homogeneous for a particular type of vegetation were selected for this purpose and yearly mown by scythe. The swath was removed immediately after mowing.

An advantage of the choice of location of the experimental areas is that these areas may be considered roughly

comparable for the abiotic environmental factors. A disadvantage, however, is the small distance between the areas under different experimental treatment. For instance, there is a chance that cattle have a preference for mown plots resulting in an increased grazing pressure as compared with the unmown ones.

In the following types of vegetation, mowing experiments were carried out. Fig. 2 shows the distribution of these types and the location of the experiments:

I. *Festuca rubra/Armeria maritima* type
II. *Artemisia maritima* type
III. *Festuca rubra/Limonium vulgare* type
IV. *Juncus maritimus* type
V. *Elytrigia pungens* (*Agropyron littorale*) type.

Prins (1976) has given a syntaxonomical and synecological characterization of the areas under the various treatments (Table 1).

Details on the species composition of the five types of vegetation can be seen from Tables 2–6.

Changes in the vegetation are followed in two ways:
– Overall changes in the study area of 32 ha will be recorded by repeated vegetation mapping. The first map was prepared for the 1971 situation, the second one for 1976. (Fig. 2 is a generalised version of the 1971-map, the detailed map is found in Prins (1976)).

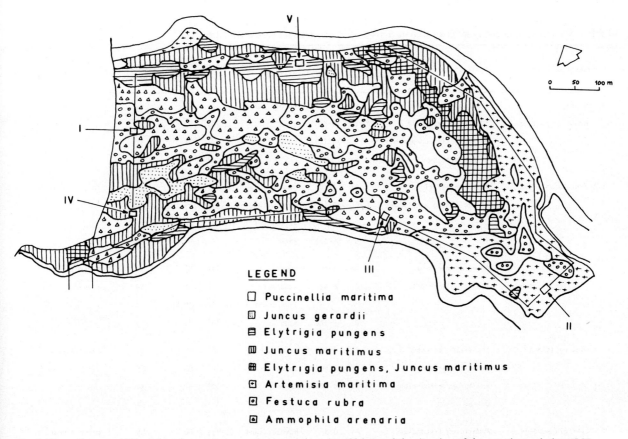

LEGEND

□ Puccinellia maritima

▣ Juncus gerardii

▤ Elytrigia pungens

▥ Juncus maritimus

▦ Elytrigia pungens, Juncus maritimus

▣ Artemisia maritima

▣ Festuca rubra

▣ Ammophila arenaria

Fig. 2. Vegetation map 1971 of the salt-marsh area under renewed grazing (32 ha) and the situation of the experimental plots (I-V).

– Changes in the experimental plots (5 × 8 m) were recorded yearly on one permanent plot (2 × 2 m) within every experimental plot. The vegetation in these plots was carefully characterised using the Braun-Blanquet approach.

Results

The vegetation map of Fig. 2 shows the whole area with a legend indicating the dominant species of the distinguished vegetation types. The *Puccinellia maritima* vegetation type is found on the lowest parts of the salt-marsh, partly along the creeks, partly in local depressions. It is surrounded by a vegetation of *Festuca rubra*, often somewhat higher situated. Farther away from the tidal flat a *Juncus gerardii* vegetation occurs in depressions with poor drainage, enclosed by low dunes. On levees bordering the tidal flat towards the salt-marsh and on creek bank levees a vegetation is found with *Artemisia maritima* as a dominant. On the highest parts of the salt marsh, low dunes have developed, covered with a vegetation of *Ammophila arenaria*. On the transition

of the latter towards lower parts of the marsh generally a vegetation of *Juncus maritimus*, *Elytrigia pungens*, or a combination of these two species is found. This arrangement of salt-marsh vegetations is found everywhere in the Netherlands (Westhoff et al. 1971). Typical for Schiermonnikoog, however, are the large areas covered by *Juncus maritimus*.

In Tables 2–6 the effects of the various experimental treatments on species composition are shown for the five types of vegetation, respectively. The columns in these tables are arranged according to the increase in overall floristic-sociological difference from the control plot.

Discussion

From the arrangement of the columns (reflecting the effect of the different treatments) resulting from the above procedure, the treatments can be divided into two groups, viz. the mowing experiments, and grazing experiments, whether or not in combination with mowing. Grazing,

99

Table 1. Syntaxonomical and synecological characterisation of the different vegetation types.

vegetation type	syntaxonomy *	soils ** profile 0-120cm		analysis 0-10cm	10-20cm	flooding frequency ***
I Festuca rubra /	Armerion maritimae	Al	0-10 : humose, sand	(1) 78	97	once a year
Armeria maritima	(Juncetum gerardii)	C	10-70 : sand	(2) 72	64	
	with brackish components	G	70-120 : sand	(3) 15	2	
	of Agropyro-Rumicion crispi			(4) 8	1	
II Artemisia maritima	Armerion maritimae	Al	0-10 : humose, clay	(1) 70	71	twice a year
	(Artemisietum maritimae)	C	10-65 : sand with clay	(2) 144	175	
			layers until 50cm	(3) -	21	
		G	65-120 : sand	(4) -	8	
III Festuca rubra/	Armerion maritimae	Al	0-16 : humose, clay	(1) 54	98	three times
Limonium vulgare	(Juncetum gerardii)	C	16-45 : sand with clay	(2) 78	60	a year
			layers until 35cm	(3) 20	1	
		G	45-120 : sand	(4) 25	1	
IV Juncus maritimus	transition between	Al	0-10 : humose, clay	(1) -	65	twice a year
	Armerion maritimae	C	10-60 : sand	(2) 95	95	
	and Agropyro-Rumicion	G	60-120 : sand	(3) -	26	
	crispi			(4) -	8	
V Elytrigia pungens	Angelicion littoralis	Al	0-8 : humose, sand	(1) 66	95	twice a year
	(Atripliceto-Elytrigietum	C	8-80 : sand	(2) 82	73	
	pungentis)	G	80-120: sand	(3) 8	4	
				(4) 26	1	

* After Westhoff & den Held (1969)
** After Prins (1976). The subsequent figures for soil analysis are: (1) sand (percentage), (2) U-figure, (3) parts $< 16\,\mu$ (percentage) and (4) organic matter (percentage)
*** After Provinciale Waterstaat Friesland. Averages over 1967–1974

especially if combined with mowing twice a year causes the most pronounced changes. In the *Elytrigia pungens* vegetation, however, all experiments cause great changes.

Tables 2–6 show great differences in the number of appearing and disappearing species. It is, however, impossible to see whether a species increases or decreases gradually or abruptly. A complete list of species of the 1971 situation for each type of vegetation has not been made. So, the question whether an appearing species is new for the vegetation type as a whole or just for one permanent plot cannot be answered. The arrangement of columns shown in Tables 2–6 is the result of the 1975 evaluation, for any other year the order may well be different.

Vegetation of Festuca rubra/Armeria maritima (Table 2)

The conclusions concerning this type of vegetation should be considered with great care because of the variation in the zero (i.e. the starting) situation as far as total cover is concerned. The control plot was very constant. This could mean that the vegetation had reached a certain equilibrium after the grazing was stopped in 1958. In nearly all experiments, except the control, total cover and height of the vegetation decreased. The *Festuca rubra* and *Potentilla anserina* cover remained constant in all treatments. Changes in cover of species present since 1971 are mainly based on an increase of the species like *Agrostis stolonifera*, *Plantago maritima*, *Poa pratensis*, *Trifolium repens*, *Carex distans*

100

Table 2. Changes in the cover of the individual species in the vegetation type of *Festuca rubra/Armeria maritima* over the 1971–1975 period.

	CONTROL	AUGUST MOWING	GRAZING	AUGUST MOWING + GRAZING	JUNE MOWING	JUNE + AUGUST MOWING	JUNE MOWING + GRAZING	JUNE + AUGUST MOWING + GRAZING
Total cover (%)	C(65)	-(60-50)	C(65)	-(50-40)	-(40-25)	C(20)	-(30-20)	-(30-15)
Height (cm)	C(20)	C(15)	-(20-2)	-(15-3)	-(25-3)	-(25-3)	-(20-3)	-(25-4)
Number of species	+(9-10)	-(11-10)	C(8)	+(9-11)	+(7-10)	+(7-10)	+(9-15)	+(9-14)
Festuca rubra	C(3)	C(2)	C(2)	C(2)	C(2)	C(2)	C(2)	C(2)
Potentilla anserina	C(2)	C(2)	C(2)	C(2)	C(2)	C(2)	C(2)	C(2)
Armeria maritima	C(1/2)	-(2-+)	C(1/2)	-(2-1)	C(2)	C(1)	C(2)	+(+-2)
Agrostis stolonifera	C(2)	C(2)	C(2)	C(1/2)	C(2)	+(0-1)1	+(+-2)	+(1-2)
Plantago maritima	C(2)	C(2)	C(2)	C(2)	C(2)	+(1-2)	+(+-2)	+(1-2)
Poa pratensis	C(+)	C(+)	C(+)	C(+)	C(1)	C(+)	+(1-2)	+(1-2)
Carex distans	C(+)	C(+)	.	+(0-+)	+(0-1)2	.	C(1)	+(0-r)4
Trifolium repens	C(1/2)	C(1/2)	C(1/2)	+(1-2)	C(2)	+(1-2)	C(1)	C(1/2)
Juncus gerardii	C(1)	+(1-2)	+(1-2)	+(1-2)	+(0-1)2	C(1)	C(+)	+(+-2)
Triglochin maritima	+(0-r)2	+(0-r)3	∧2	∧2	.	.	+(0-r)4	+(0-+)1
Artemisia maritima	.	-(1-0)1	∧3	.	.	∧3	+(0-r)3	C(+)
Atriplex hastata	.	-(+-0)1	.	-(+-0)1	.	∧3	.	.
Spergularia media	.	.	∧2,3	+(0-r)4	.	.	.	.
Centaurium littorale	.	.	.	+(0-1)4	.	.	.	.
Lotus corniculatus	.	.	.	.	.	+(0-+)2	+(0-+)3	+(0-r)4
Sagina maritima	.	.	.	.	.	+(0-r)3	+(0-+)4	.
Plantago coronopus	.	.	.	.	+(0-r)2	.	+(0-r)4	+(0-r)4
Cochlearia danica	.	.	.	.	.	.	+(0-r)4	.
Elytrigia pungens	.	.	.	.	.	.	.	+(0-1)1

Symbols: C: constant; .: does not occur; +(+ –3): increase + to 3 according to the Braun-Blanquet scale; −(1–r): decrease 1 to r after Br.-Bl.; +(0–1)2: species establishing in the second year of the experiment, cover 1 in 1975 after Br.-Bl.; ∧ 3: species appearing and disappearing in the third year of the experiment.

Table 3. As Table 2, vegetation type of *Artemisia maritima*

	CONTROL	JUNE MOWING	JUNE + AUGUST MOWING	AUGUST MOWING	JUNE MOWING + GRAZING	JUNE + AUGUST MOWING + GRAZING	GRAZING	AUGUST MOWING + GRAZING
Total cover (%)	C(90)	C(80)	C(80)	+(50-90)	-(80-50)	-(80-60)	C(90)	-(90-70)
Height (cm)	C(30)	-(30-20)	-(30-10)	C(20)	-(20-5)	-(20-5)	-(30-20)	-(20-5)
Number of species	+(8-9)	-(9-8)	+(7-8)	-(9-8)	C(9)	+(6-12)	+(9-13)	+(9-15)
Artemisia maritima	C(4)	C(4)	C(3/2)	+(2-3)	-(3-2)	-(3-2)	-(4-3)	-(4-2)
Festuca rubra	C(2)	C(2)	+(2-3)	+(2-3)	C(2)	C(2)	C(2)	-(3-2)
Juncus gerardii	C(1)	-(1-+)	+(1-2)	+(1-2)	+(1-2)	+(1-2)	+(1-2)	+(1-2)
Plantago maritima	C(2)	C(1)	C(2)	C(2)	C(1/2)	C(2)	C(2)	+(1-2)
Limonium vulgare	C(+)	C(+)	C(+)	C(2)	C(+)	C(1)	C(1)	-(1-+)
Atriplex hastata	C(+)	C(+)	C(+)	C(+)	.	.	C(+)	C(+)
Agrostis stolonifera	+(0-+)2	+(+-1)	+(0-r)4	+(+-2)	+(0-1)1	+(0-1)1	+(+-2)	+(1-2)
Glaux maritima	C(+)	-(+-0)1	.	-(+-0)4	C(+)	+(+-2)	C(1)	+(0-+)2
Triglochin maritima	-(2-+)	C(+)	+(+-2)	C(1)	+(1-2)	+(1-2)	+(+-1)	C(2)
Armeria maritima	.	.	.	.	-(+-0)1	+(0-+)2	+(0-r)4	+(0-+)3
Spergularia media	.	.	.	.	.	.	-(+-0)4	C(+)
Salicornia europaea	.	.	.	.	.	+(0-r)2	+(0-r)4	+(0-+)3
Aster tripolium	.	.	.	.	C(r)	C(+)	.	+(0-+)3
Suaeda maritima	.	.	.	.	.	+(0-r)2	.	+(0-r)3
Puccinellia maritima	.	.	.	.	.	.	+(0-r)4	.
Spergularia marina	.	.	.	.	.	.	+(0-r)4	+(0-r)3

Table 4. As Table 2, vegetation type of *Festuca rubra/Limonium vulgare*

	CONTROL	AUGUST MOWING	JUNE MOWING	JUNE + AUGUST MOWING	AUGUST MOWING + GRAZING	JUNE MOWING + GRAZING	GRAZING	JUNE + AUGUST MOWING + GRAZING
Total cover (%)	+(60-100)	+(60-100)	C(75)	C(80)	C(60)	C(60)	C(75)	-(70-50)
Height (cm)	C(30)	-(40-20)	-(30-15)	-(40-10)	-(20-10)	-(20-5)	-(25-10)	-(20-5)
Number of species	-(9-8)	+(8-9)	C(6)	C(7)	+(9-15)	+(6-10)	+(8-13)	+(7-11)
Festuca rubra	+(3-5)	+(3-4)	C(4)	C(5)	C(2)	-(4-3)	+(2-4)	+(3-2)
Limonium vulgare	+(1-2)	+(1-2)	.	C(1)	C(1)	C(1)	+(1-2)	+(1-2)
Artemisia maritima	+(1-2)	C(1)	C(+)	C(+)	C(1)	C(1)	-(2-1)	C(1/2)
Juncus gerardii	C(+)	C(1)	.	C(1)	C(2)	.	C(2)	.
Plantago maritima	C(1)	C(1)	+(1-2)	C(1)	C(2)	+(1-2)	+(1-2)	+(1-2)
Agrostis stolonifera	-(2-1)	+(1-2)	+(1-2)	+(1-2)	C(1/2)	+(1-2)	+(1-2)	+(1-2)
Atriplex hastata	-(+-r)	+(0-1)1	.	C(1)	C(+)	+(0-+)3	+(0-+)1	.
Glaux maritima	-(+-0)4	-(+-0)3	.	.	C(1)	C(1)	C(1)	+(+-1)
Triglochin maritima	-(+-r)	C(1)	.	.	C(1)	.	C(1)	+(0-+)1
Armeria maritima	.	+(0-+)1	C(+)	+(r-1)	+(0-+)1	.	.	+(r-1)
Trifolium repens	.	.	+(r-1)	.	+(0-r)4	+(0-r)4	.	.
Salicornia europaea	.	.	.	.	+(0-r)4	+(0-r)3	+(0-r)2	+(0-r)3
Odontites verna	.	.	.	.	+(0-r)4	.	.	.
Cochlearia anglica	.	.	.	.	+(0-r)4	.	.	.
Spergularia media	.	.	.	.	+(0-r)4	.	+(0-+)2	.
Puccinellia maritima	.	.	.	.	.	+(0-r)4	+(0-r)4	.
Spergularia marina	.	.	.	.	.	.	+(0-r)4	.
Suaeda maritima	.	.	.	.	.	.	.	+(0-r)3
Aster tripolium	.	.	.	.	.	.	.	+(0-r)3

Table 5. As Table 2, vegetation type of *Juncus maritimus*

	CONTROL	JUNE + AUGUST MOWING	JUNE MOWING + GRAZING	JUNE MOWING	AUGUST MOWING	JUNE + AUGUST MOWING + GRAZING	GRAZING	AUGUST MOWING + GRAZING
Total cover (%)	+(75-100)	C(40)	C(35)	C(40)	+(50-90)	C(35)	C(60)	C(60)
Height (cm)	C(60)	-(60-15)	-(40-5)	-(60-40)	C(40)	-(40-5)	-(60-40)	-(30-3)
Number of species	C(6)	+(7-9)	+(8-10)	+(6-11)	+(8-14)	+(7-14)	+(6-11)	+(9-12)
Juncus maritimus	C(3)	C(2)	C(2)	C(2)	C(2)	C(2)	C(2)	-(2-1)
Festuca rubra	C(2/3)	C(2)	C(2)	C(2)	+(3-5)	C(2)	C(2)	C(2)
Atriplex hastata	C(+)	.	C(+)	C(1)	C(1)	C(+)	C(1)	-(1-r)
Agrostis stolonifera	+(+-3)	+(+-2)	+(1-2)	+(1-2)	C(2)	+(+-2)	+(+-2)	C(2)
Juncus gerardii	C(1)	C(1)	.	C(1)	C(+)	+(1-2)	+(1-2)	+(1-2)
Glaux maritima	-(2-+)	C(+)	+(+-2)	C(1)	C(1/2)	+(+-2)	C(2)	+(1-2)
Artemisia maritima	.	+(0-+)3	+(r-+)	+(0-r)4	-(1-r)	+(+-1)	+(0-+)3	+(0-r)2
Potentilla anserina	.	+(1-2)	.	+(0-+)2	+(0-1)3	+(0-+)3	.	.
Poa pratensis	.	+(0-+)3	.∗	+(0-1)2	+(0-+)4	+(0-r)4	.	.
Cerastium holosteoides	.	.	.	+(0-r)4	.	.	.	.
Salicornia europeae	.	.	+(0-+)3	.	.	.	.	.
Cirsium arvense	.	.	.	+(0-r)4	.	.	.	.
Trifolium repens	.	+(r-+)	.	.	+(0-+)3	+(0-+)3	.	.
Spergularia media	.	.	+(0-+)3	.	.	+(0-r)2	.	+(0-+)3
Leontodon autumnalis	.	.	.	.	+(0-r)3	.	.	.
Plantago maritima	.	.	C(1)	.	+(0-r)3	.	+(0-+)3	C(+).
Armeria maritima	.	.	.	.	+(0-r)3	+(0-1)2	.	-(+-0)3
Triglochin maritima	.	.	.	.	C(+)	+(0-1)2	.	+(1-2)
Puccinellia maritima	.	.	.	.	.	+(0-r)3	+(0-1)4	+(0-+)4
Spergularia marina	.	.	.	.	.	.	+(0-1)3	.
Suaeda maritima	.	.	.	.	.	.	+(0-r)3	.
Parapholis strigosa	.	.	.	.	.	.	.	+(0-r)4

Table 6. As Table 2, vegetation type of *Elytrigia pungens*

	CONTROL	AUGUST MOWING	AUGUST MOWING + GRAZING	GRAZING	JUNE MOWING	JUNE + AUGUST MOWING	JUNE MOWING + GRAZING	JUNE + AUGUST MOWING + GRAZING
Total cover (%)	+(30-60)	+(25-50)	C(30)	+(30-50)	+(30-50)	C(40)	C(30)	C(30)
Height (cm)	C(60)	C(50)	-(30-5)	-(60-10)	-(50-20)	-(50-10)	-(50-10)	-(50-5)
Number of species	+(5-6)	+(5-15)	+(7-15)	+(4-12)	+(3-11)	+(4-14)	+(4-14)	+(5-15)
Elytrigia pungens	+(3-4)	C(2/3)	C(2)	-(3-2)	-(3-2)	-(3-2)	-(3-2)	-(3-1)
Festuca rubra	+(+-1)	+(+-2)	+(1-2)	+(+-2)	+(1-2)	+(1-2)	C(+)	+(1-2)
Artemisia maritima	+(0-r)4	+(+-1)	+(+-1)	+(0-+)3	+(0-1)1	+(0-1)1	+(0-1)1	+(0-1)1
Atriplex hastata	C(+/1)	C(+)	C(+)	C(+)	.	C(1)	C(+)	C(+)
Juncus maritimus	C(+)	+(0-+)2	C(+)	C(+)	+(+-1)	+(+-1)	+(+-2)	C(1)
Agrostis stolonifera	C(+)	+(+-2)	+(0-2)1	+(0-1)2	+(0-+)1	+(0-2)1	+(0-2)1	+(r-2)
Cerastium holosteoïdes	.	+(0-1)2	+(0-1)1	+(0-r)4	+(0-+)1	+(0-1)2	+(0-2)1	+(0-1)1
Carex distans	.	+(0-+)2	+(0-2)1	+(0-r)4	+(0-r)1	+(0-+)2	+(0-2)2	+(0-1)1
Trifolium repens	.	+(0-1)2	+(0-1)1	+(0-r)4	+(0-1)1	+(0-r)3	+(0-1)2	+(0-1)1
Poa pratensis	.	+(0-1)2	+(0-1)1	.	+(0-r)3	+(0-1)2	+(0-+)1	+(0-2)1
Sagina maritima	.	+(0-+)2	+(0-+)3	+(0-r)4	.	.	+(0-1)2	+(0-2)1
Armeria maritima	.	+(0-r)2	.	.	.	.	.	.
Parapholis strigosa	.	+(0-r)4	.	.	.	.	.	.
Lotus corniculatus	.	+(0-r)2	.	.	.	.	.	.
Sonchus oleraceus	.	∧1	.	.	.	.	.	.
Glaux maritima	.	∧2	.	.	.	.	.	.
Leontodon autumnalis	.	∧1,3	.	.	.	.	.	.
Plantago maritima	.	+(0-+)2	+(0-r)3	.	.	.	.	.
Limonium vulgare	.	∧2	-(+-0)1	.	.	+(0-r)1	.	.
Cirsium arvense	.	∧1	-(+-0)1	.	.	.	.	.
Spergularia marina	.	.	+(0-r)3	+(0-r)4	.	.	.	.
Potentilla anserina	.	∧1,2	+(0-+)1	.	+(0-r)3	.	.	.
Plantago coronopus	.	.	+(0-+)3	.	+(0-r)2	+(0-r)3	.	.
Puccinellia maritima	.	.	.	+(0-r)4	.	+(0-+)2	.	.
Bupleurum tenuissimum	.	∧1,3	.	.	.	+(0-1)1	+(0-r)3	+(0-1)2
Stellaria graminea	.	∧1	.	.	.	.	+(0-r)3	+(0-+)3
Cochlearia danica	.	.	.	.	.	.	+(0-+)2	+(0-+)2
Trifolium fragiferum	.	.	.	.	.	.	.	+(0-+)3

and *Juncus gerardii*. The *Armeria maritima* cover decreased in some treatments.

Syntaxonomically the vegetation type remained a transition between *Armerion maritimae* and *Agropyro-Rumicion crispi*. The new appearance of *Sagina maritima, Plantago coronopus, Cochlearia danica* and, into a lesser extent, *Centaurium littorale* indicates a shift in the direction of the *Saginion maritimae* and is presumably due to a decrease in the total cover.

The treatments causing the largest change appear to have one factor in common, namely the mowing of the vegetation early in the growing season, followed either by mowing again, or grazing, in other words cultural practices which keep the vegetation short.

The observed changes for the species are fairly equally distributed over the period of observation. However, there is some evidence that the treatments with the largest effects show these effects particularly in the last year of observation.

Vegetation of Artemisia maritima (Table 3)

The control plot was very constant indicating that the equilibrium situation had been reached in 1971 or earlier, after grazing was stopped in 1958. Height of the vegetation decreased in all experiments except in the control, while total cover decreased especially in the late mowing and/or grazing experiments. These decreases were closely linked with the decrease of *Artemisia maritima*, the dominant species in this vegetation type. This could be expected because the species is fully developed only in late summer. The cover of *Plantago maritima, Limonium vulgare* and *Atriplex hastata* remained constant in almost all experiments. The cover of *Festuca rubra* increased in case of mowing late in the summer in the absence of grazing.

Most experiments showed an increase of *Juncus gerardii* and *Agrostis stolonifera*. Together with the increase of *Armeria maritima* and *Triglochin maritima* in nearly all grazed plots, this could indicate a shift from the *Artemisietum maritimae* to the *Juncetum gerardii* within the *Armerion maritimae*. The appearance of *Spergularia marina* and

Puccinellia maritima in a number of grazed plots could mean a change towards the *Puccinellio-Spergularion maritimae* and, therefore, an indication of a regression in the halosere. This is supported by the new appearance of *Salicornia europaea* and *Suaeda maritima*. These changes are presumably due to the decrease of the dominant species and trampling of the sod by cattle.

The treatments showing the greatest effect have one factor in common, viz. grazing whether or not in combination with late mowing.

The greatest change had taken place during the last years of the experiments. It is remarkable to see that four new species appeared in the fourth year of the grazing treatment. Five new species appeared in the third year of the combined treatment of grazing and mowing in August.

Vegetation of Festuca rubra/Limonium vulgare (Table 4)

The control plot was not constant. Since 1971 the vegetation developed into the direction of a *Festuca rubra* dominance at the cost of other species e.g. *Agrostis stolonifera*, *Glaux maritima* and *Triglochin maritima*. It is not clear whether this phenomenon has to do with succession or with the stopping of grazing in 1958. Human interference seems necessary to reach a certain vegetation equilibrium viz. mowing in June and August or grazing and mowing in August. This at least was the impression so far after five years of treatments. The height of the vegetation decreased under all treatments, except in the control. Total cover increased in the control and in the August mowing plot. This is probably caused mainly by the increase of the *Festuca rubra* cover. Only in the most intensive treatment viz. grazing and twice mowing, total cover decreased. In most experiments *Agrostis stolonifera* increased while *Artemisia maritima* and *Juncus gerardii* remained constant. *Plantago maritima* especially increased in the grazing experiments.

Syntaxonomically, this vegetation type is an *Armerion maritimae* vegetation which is difficult to subdivide. The appearance of *Trifolium repens* and *Spergularia media* under a number of grazing treatments indicates a development towards a *Juncetum gerardii*. In the same permanent plots the new appearance of *Salicornia europaea*, *Puccinellia maritima* and *Cochlearia anglica* (*Puccinellion maritimae* species) is most likely induced mainly by trampling of the sod by cattle.

The experiments showing the greatest changes had the factor grazing in common. The largest changes had been found during the last years of the experiments. It is re-

markable that in the combined grazing and August mowing experiment, five new species appeared suddenly.

Vegetation of Juncus maritimus (Table 5)

The control plot was rather constant. A striking increase of *Agrostis stolonifera* at the cost of *Glaux maritima* was observed. *Juncus maritimus* remained constant in nearly all experiments. A decrease for this species was only found under grazing combined with August mowing. *Festuca rubra* only showed a great increase in the August mowing treatment. *Agrostis stolonifera* increased in nearly all treatments and the same was seen for *Artemisia maritima*. Total cover remained constant in most of the experiments. Height of vegetation decreased in all experiments, except in the control plot.

In most of the mowing treatments the *Agropyro-Rumicion crispi* species *Potentilla anserina*, *Trifolium repens* and *Poa pratensis* appeared new. Under most of the grazing treatments, however, *Glaux maritima*, *Spergularia media* and *Triglochin maritima* either appeared new or increased their cover, which indicates a higher number in elements of the *Armerion maritimae*. The new appearance of *Puccinellia maritima* in nearly all, and *Salicornia europaea* and *Parapholis strigosa* under some of the grazing treatments indicates a *Puccinellion maritimae* element, probably caused by trampling of the sod by cattle.

Grazing, mowing, as well as grazing combined with mowing, may cause large changes here. Changes took place especially in the third year of the experiments, particularly with the mowing-in-August and grazing-only treatments. In the other years hardly any changes were observed. It is remarkable that no new species at all were found after the first year of treatment, in contrast with the other four vegetation types.

Vegetation of Elytrigia pungens (Table 6)

The control plot was fairly constant. Since 1971, the greatest change was the increase of *Elytrigia pungens*. This is probably due to the fact that grazing ceased in 1958. Especially in the early mowing treatments, whether or not in combination with grazing, *Elytrigia pungens* decreased. Total cover remained constant in the grazing treatments in combination with mowing. In the other treatments, however, it increased. A decrease in the height of vegetation was seen under almost all treatments, but not for the control plot. Both *Festuca rubra* and *Artemisia maritima*, in the control plot as well as under all other treatments, showed an increase in cover.

The increase in cover or new appearance of *Agrostis stolonifera*, *Cerastium holosteoides*, *Carex distans*, *Trifolium repens* and *Poa pratensis* under all experiments, except in the control one, is remarkable. This indicates a change to the *Agropyro-Rumicion crispi*. However, the new appearance of *Sagina maritima*, *Plantago coronopus*, *Bupleurum tenuissimum* and *Cochlearia danica* is an indication of a shift towards the *Saginion maritimae*. The latter predominantly occurred under the grazing treatments suggesting a probable influence of trampling of the sod by cattle.

All treatments had large effects. This is most probably due to the fact that in all treatments the thick layer of mostly dead *Elytrigia pungens* material vanished.

The changes in this vegetation type took place especially in the first two years of the treatments. Under the grazing treatment, however, the changes were only found in the fourth year of the treatment.

Some preliminary conclusions

From the data discussed above, some preliminary conclusions can be drawn:
– Under the different management experiments, two types of changes took place. The vegetation types of *Festuca rubra/Armeria maritima* and *Elytrigia pungens* showed, establishment and increase of *Saginion maritimae* elements, indicating a greater openness of the vegetation. The vegetation types of *Festuca rubra/Limonium vulgare* and *Juncus maritimus* on the one side and *Artemisia maritima*

Table 7. General syntaxonomical changes observed over the period 1971–1975 under the influence of the various grazing and mowing experiments.

vegetation type (dominant species)	most important syntaxonomical element in 1971	syntaxonomical elements newly appeared since 1971
Festuca rubra/ Armeria maritima	Armerion maritimae/ Agropyro-Rumicion crispi	Saginion maritimae
Artemisia maritima	Armerion maritimae	Puccinellio-Spergularion maritimae
Festuca rubra/ Limonium vulgare	Armerion maritimae	Puccinellion maritimae
Juncus maritimus	Armerion maritimae/ Agropyro-Rumicion crispi	Puccinellion maritimae
Elytrigia pungens	Angelicion littoralis	Saginion maritimae

Table 8. Relative degree of changes in the different vegetation types under the influence of the various grazing and mowing experiments observed over the period 1971–1975.

vegetation type (dominant species)	control	August mowing	June mowing	June + August mowing	grazing	August mowing + grazing	June mowing + grazing	June + August mowing + grazing
Festuca rubra/ Armeria maritima	--	--	--	--	--	-	-	-
Artemisia maritima	--	-	--	-	+	+	-	+
Festuca rubra/ Limonium vulgare	+	-	--	--	+	+	+	+
Juncus maritimus	-	+	-	-	+	+	-	+
Elytrigia pungens	-	++	++	++	++	++	++	++

Symbols: + +: large change; +: rather large change; −: rather small change; −−: small change.

on the other one, showed a tendency to regression expressed in the development of *Puccinellion maritimae* and *Puccinellio-Spergularion maritimae* elements, respectively (Table 7).

– The changes caused by the different treatments varied, which is qualitatively shown in Table 8. Changes were small in the *Festuca rubra/Armeria maritima* vegetation: they only occurred under the treatments grazing in combination with mowing. In the vegetation types of *Artemisia maritima*, *Festuca rubra/Limonium vulgare* and *Juncus maritimus* changes were rather large especially under grazing treatments.

Changes in the *Elytrigia pungens* vegetation were large under all treatments. Special attention is needed for the control plots of all vegetation types. They show the relative changes observed over a five-year period (1971–1975), and therefore give some idea of the equilibrium reached 13 years after grazing had ceased (1958–1971). The vegetation types of *Festuca rubra/Armeria maritima* and *Artemisia maritima* did not change anymore. The vegetation types of *Juncus maritimus* and *Elytrigia pungens* did not seem to be quite equilibrated, whereas the vegetation of *Festuca rubra/Limonium vulgare*, still changed considerably.

– The treatments causing the most important changes (Cf. Table 8) varied in their rate of change in the different types of vegetation (Table 9). Changes were gradual in the *Festuca rubra/Armeria maritima* vegetation. In the vegetation types of *Artemisia maritima*, *Festuca rubra/Limonium vulgare* and *Juncus maritimus* changes were rather abrupt, taking place especially in the third and fourth year of the

Table 9. Rate of the most important changes in the different vegetation types under the influence of the grazing and mowing experiments observed over the period 1971–1975.

vegetation type (dominant species)	rate of change
Festuca rubra/Armeria maritima	gradual
Artemisia maritima	abrupt, third and fourth year
Festuca rubra/Limonium vulgare	abrupt, third and fourth year
Juncus maritimus	abrupt, third and fourth year
Elytrigia pungens	abrupt, first and second year

experiments. Changes were also abrupt in the *Elytrigia pungens* vegetation but often found in the first and second year of the experiments. Except in the *Festuca rubra/Armeria maritima* vegetation, possibilities for settlement for more species suddenly arose simultaneously, not (yet) being established in previous years.

It is virtually impossible to evaluate and generalise the observations above described for a comparison with data from elsewhere. Although a wealth of literature exists on salt-marsh vegetation and on the effect of grazing in such areas, very little is known on how changes actually take place and when an equilibrium is reached between human interference and nature's response. Almost all literature deals with a one time comparison of grazed and ungrazed areas. The data presented above are only the first steps in following the process of the impact of various management regimes on a salt-marsh vegetation. Long term observations are a prerequisite to observe it and in a later stage long term effects may be predicted. The present study is continued. The method applied can possibly instigate investigations on the population dynamics and autecology of the participating species.

Summary

Part of a salt-marsh (32 ha), ungrazed from 1958 until 1971, was grazed again from 1972 onwards with young cattle (1.3–1.7 per ha, May-October). In five vegetation types management experiments, including doing nothing (control), June mowing, August mowing, June and August mowing, all in combination or not in combination with grazing have been started with the objective to compare annually their effects on the vegetational structure and composition by means of permanent plots (2 × 2 m).

Thirteen years (1958–1971) after the grazing had ceased the vegetation types of *Festuca rubra/Armeria maritima*, *Artemisia maritima*, *Juncus maritimus* and *Elytrigia pungens* hardly changed anymore. The vegetation type of *Festuca rubra/Limonium vulgare*, changed considerably. The experiments showed rather specific effects during the period 1971–1975 for each type of vegetation. Changes in the *Festuca rubra/Armeria maritima* vegetation were small and gradually under all treatments. Changes in the *Artemisia maritima*, *Festuca rubra/Limonium vulgare* and *Juncus maritimus* vegetation types were rather great under the different grazing treatments, the changes being abrupt and especially taking place in the third and fourth year of the experiments. The *Elytrigia pungens* vegetation showed large changes under all treatments, except the control plot, whereas these changes were abrupt, particularly occurring in the first and second year of the experiments. The study is continued.

References

Beeftink, W.G. 1977. Salt-marshes. In: Handbook on the applied ecology and management of the coastline. Ed. R.S.K. Barness. Wiley, Chichester.

Bleuten, W. 1971. Een geomorfologische studie van het eiland Schiermonnikoog. Internal report, State University, Utrecht (mimeo) 70 pp.

Hartog, C. den. 1952. Plantensociologische waarnemingen op Schiermonnikoog. Kruipnieuws 14(2): 2–24.

Heukels, H. & S.J. van Ooststroom. 1970. Flora van Nederland. Wolters-Noordhoff, Groningen. 909 pp.

Oosterveld, P. 1975. Beheer en ontwikkeling van natuurreservaten door begrazing. Natuur Landschap 29(6): 161–171.

Prins, F.W. 1976. Beweiding op de Oosterkwelder, Schiermonnikoog. Internal report, State University, Groningen/ Research Institute for Nature Management, Leersum (mimeo), 36 pp.

Westhoff, V. 1954. Landschap en plantengroei van Schiermonnikoog. Natuur Techniek 22 (5/6): 1–10.

Westhoff, V. & A.J. den Held. 1969. Plantengemeenschappen in Nederland. Thieme, Zutphen. 324 pp.

Westhoff, V., P.A. Bakker, C.G. van Leeuwen & E.E. van der Voo. 1971. Wilde Planten deel 1. De Lange/Van Leer, Deventer. 320 pp.

DYNAMICS OF BENTHIC ALGAL VEGETATION AND ENVIRONMENT IN DUTCH ESTUARINE SALT MARSHES, STUDIED BY MEANS OF PERMANENT QUADRATS*·**

P.H. NIENHUIS

Delta Institute for Hydrobiological Research, Yerseke, The Netherlands***

Keywords: Netherlands, Permanent quadrats, Salt-marsh algae, Salt-marsh environment, Vegetation dynamics

Introduction

Mud flats and salt marshes under tidal influence and at the borders of non-tidal brackish inland waters form characteristic semi-natural landscapes along the coast of the Netherlands. Most botanical studies done in these saline environments deal with higher plants (Beeftink 1977). However, besides the phanerogams, benthic algae take an important place in these ecosystems. The algae concerned are in general multicellular plants greatly varying in size (from 1–2 microns in diameter up to a thallus length of over 1 m). These algae live upon and in the uppermost centimetres of the soil, as epiphytes, or entangled with phanerogams, or floating in shallow pools.

Knowledge of the spatial structure of benthic algal vegetations in salt marshes consisting of about 100 species of blue-green, green, brown and red algae, formed the basis of a study into the dynamics of the algal mat (Nienhuis 1969, 1970, 1975). The temporal changes in algal vegetations and concomitant processes in their environments were studied in 27 permanent sample plots, established in tidal marshes and along brackish inland waters in the SW Netherlands. A detailed analysis of these data was published by Nienhuis (1975). The aim of this paper is to evaluate and to summarize the results.

Material and methods

To study the temporal changes occurring in the algal vegetation on soft substrates permanent quadrats (PQs) were used. The size of the PQ must at any rate be larger than that of the minimum area of the vegetation. The PQ has to cover a surveyable part of a homogeneous vegetation. My PQs measured 2 m × 2 m each. In the centre of the PQ a square of 0.5 × 0.5 m was considered to be the PQ proper: it was left undisturbed during the investigations. The fringe served as a sampling area for algae and soil. A permanent wooden picket, protruding a few cm above soil-level, marked the centre of the PQ, which was delimited by a loose 0.5 m × 0.5 m iron frame each time it was studied.

Once a month a relevé of the PQ proper was made. From each species or species group that could be distinguished in the PQ proper, a sample was taken from the fringe area; this sample was supposed to give a reliable picture of the PQ proper.

The degree of similarity between the relevés sampled in the course of time was expressed by the index of Sörensen (1948):

$$Q_s = \frac{2c}{a+b} \times 100$$

Q_{sm} is the average degree of similarity over a period of one year. A high or low value of Qsm indicates small, respectively large changes in the species composition of the PQ in the course of one year. The number of algal taxa (N) per PQ per month is taken as a measure for species diversity. Nm means the average number of taxa per PQ over a period of one year.

The analysis of a permanent quadrat comprised the following steps:

* Nomenclature authorities are given in Table 3.
** Contribution to the Symposium of the Working Group for Succession Research on Permanent Plots, held at Yerseke, The Netherlands, October 1975.
*** Communication No. 163 of the Delta Institute for Hydrobiological Research, Yerseke.

1. a short general description of the vegetation and its environment;
2. estimate of the percentage coverage of each of the main plant groups (flowering plants, mosses, lichens, algae) in the quadrat, in each vegetation layer; estimate of the combined abundance-cover degree of each algal species, or species group in the case of algal mats, using the phytosociological scale of Doing Kraft (1954);
3. sampling of the algae for further microscopical analysis in the laboratory;
4. correction of the field relevé as regards the share of the algal species in the samples found through microscopical analysis.

The abundance-cover degree scale of Doing Kraft (1954), slightly modified to my purpose, and used for field analysis, was as follows:

O = cover less than 5 %, individuals rare or occasional
F = cover less than 5 %, individuals frequent
01 = cover 5–15 %, number of individuals arbitrary
02 = cover 15–25 %, etc., up to 10 = cover 95–100 %, number of individuals arbitrary.

No estimates were made of the sociability of the algal vegetation, because in most cases it is hardly possible to distinguish individual algal plants.

It is impossible to estimate coverage of individual algal species forming algal mats or turfs in the field with the degree of precision of the above scale. Moreover, most species can only be recognized with the aid of a microscope. Therefore the following procedure was used to obtain rough estimates of the abundance-cover degree of algal species:

1. A representative sample of 4 to 5 cm^2 was collected from each algal species or species group as recognized in the field; it was cut from the algal mat or scraped from the substrate with a knife.
2. Each of these intact samples was inspected with a low power binocular stereo microscope and, if necessary, divided into subsamples of thus recognizable species or species groups; the cover degrees noted in the field were corrected accordingly.
3. At least two microscopic slides were prepared from each of these samples or subsamples for inspection under the high power microscope. These two slides, for which large coverslips of 2.4 × 3.2 cm were used, contained the greater part of the algal material of the sample or subsample. They were carefully examined and the abundance-cover degree of each species was estimated, using the following rough scale:
O = cover less than 5%, individuals rare or occasional
F = cover less than 5%, individuals frequent

C = cover 5–30%, number of individuals arbitrary, often common
A = cover 30–60%, number of individuals arbitrary, often abundant
D = cover 60–100%, number of individuals arbitrary, often dominant

After that, the mean abundance-cover degree for the two slides was determined.
4. The thus obtained mean abundance-cover degree of each species on the slides was multiplied with the abundance-cover degree of the species in the field (both values as means of their ranges) and divided by 100. This gives the abundance-cover degree of an individual species as participant of one species group. The values a particular species gained in all species groups were added and so, the combined abundance-cover degree of that species in the whole field relevé was arrived at.

The minimum area size of a sample was tested (Nienhuis 1975): A sample is sufficiently representative when more than 80 % of the species occurring in a homogeneous part of the vegetation can be found in that sample.

Once a month vegetation and environment of the PQs were examined over the period March 1968 up to and including February 1971. The data on the algal vegetation were correlated with environmental parameters viz. the moisture content of the upper centimetre of the soil and the salinity of the soil moisture, and with the cover percentage of the accompanying phanerogams. Data on tidal oscillations and water-level fluctuations, precipitation and evaporation were collected (cf. Nienhuis 1975).

The following hydrographical terminology is used in this paper. In tidal waters: MHW = mean high water; MHWS = mean high water of springtides; EHWS = extreme high water of springtides. In non-tidal waters: EHWL = extreme high water level.

In this study aquatic, semi-terrestrial and terrestrial environments are considered. Under tidal conditions an aquatic environment is defined as an environment with a flooding frequency of more than 30 % and approximately coinciding with the area below MHWS. A semi-terrestrial environment is defined as an environment with a flooding frequency of 5–30 % and approximately coinciding with the area from MHWS up to just below EHWS. A terrestrial environment is defined as an environment with a flooding frequency of less than 5 % and approximately coinciding with the area higher than a level just below EHWS.

Under non-tidal conditions an aquatic environment is defined as an environment which is inundated most of the time, whereas only the upper margin emerges several times a year. A semi-terrestrial environment is defined as an environment which is inundated several times a year. A terrestrial environment is defined as an environment of which only the lower margin submerges several times a year, coinciding with the area higher than a level just below EHWL (Nienhuis 1975).

The areas

Permanent quadrats were established in the following saline tidal localities:
(1) Springersgors (Sp), an ungrazed salt-marsh area in the Isle of Goeree-Overflakkee, in the Grevelingen estuary, near the North-Sea coast (18 PQs).

Table 1. Ranking of parameters studied in permanent quadrats in the SW Netherlands.
Aq = aquatic, Te = semi-terrestrial and terrestrial position of the PQ;
Qsm = average similarity; 5 = 80-100, 4 = 60-80, 3 = 40-60, 2 = 20-40,
1 = 0-20; Nm = average number of species: 5 = 8->10, 4 = 6-8, 3 = 4-6,
2 = 2-4, 1 = 0-2; A = fluctuations in soil moisture content and salinity of the soil moisture: 3 = large, 2 = moderate, 1 = small; B = stability of the substrate: 3 = large, 2 = moderate, 1 = little; C = maximum
cover percentage of phanerogams: 3 = 70-100%, 2 = 30-70%, 1 = less than 30%.

Aq/Te	PQ		March 1968-February 1969					March 1969-February 1970					March 1970-February 1971				
			Qsm	Nm	A	B	C	Qsm	Nm	A	B	C	Qsm	Nm	A	B	C
Aq	Sp	2	4	5	2	2	3	3	4	2	2	3	4	5	3	2	3
Aq	Sp	3	3	4	1	2	3	3	3	1	2	3	3	4	1	2	3
Aq	Sp	4	4	4	2	1	2	3	2	2	1	2	4	3	2	1	3
Aq	Sp	5	3	3	1	2	3										
Aq	Sp	6	2	2	1	2	3										
Aq	Sp	7	4	3	2	2	2										
Aq	Sp	9	3	5	1	2	3										
Aq	Sp	10	1	2	1	2	3										
Aq	Sp	11	1	1	2	2	3										
Te	Sp	13	4	5	3	3	2	4	5	3	3	2	5	5	3	3	2
Aq	Sp	15	3	4	2	2	3	3	3	2	2	3					
Te	Sp	16	4	5	3	3	2	4	5	3	3	2					
Te	Sp	18	3	4	3	2	1	1	2	3	2	1					
Te	Sp	19	3	4	.3	3	2	3	3	3	3	2					
Aq	Sp	20	4	3	1	2	3	4	3	1	2	3					
Aq	Sp	21	3	3	1	1	3	3	3	1	1	2					
Aq	Sp	22						4	4	3	2	1	5	4	3	2	1
Aq	Sp	23						3	4	2	1	1	4	4	3	1	2
Te	Bat	1						4	5	3	3	2	5	5	3	3	2
Te	Bat	2						4	5	3	3	2	5	5	3	3	2
Te	Suz	1						4	5	3	3	3	4	4	3	3	3
Te	Suz	2						3	2	3	1	1	4	2	3	1	1
Te	Via	1						4	5	3	3	3					
Te	Via	2						4	5	3	3	3					
Te	Via	3						3	3	3	1	1					
Te	DW	1						4	5	3	2	2	4	4	3	2	2
Te	DW	2						4	4	3	1	1	4	3	3	1	2

(2) Salt marsh near Battenoord (Bat), in the Isle of Goeree-Overflakkee as well in the Grevelingen estuary. The marsh is used as pasture for sheep grazing (2 PQs).

At the border of the following non-tidal brackish inland water-bodies PQs were established:

(3) Suzanna's Inlaag (Suz), a low-lying brackish area enclosed by an outer and an inner dike, and bordering the Oosterschelde. The area is grazed by cattle and sheep (2 PQs).

(4) Deesche Watergang (DW), a brackish watercourse fringed by low-lying swampy grassland. The pastures are grazed by cattle, horses and sheep during the dry seasons (2 PQs).

(5) Spuikom Vianen (Via), a shallow brackish pond near the Oosterschelde. The banks are weak and marshy; they consist of peat and bear a dense vegetation of higher plants and algae. The area is grazed by sheep except the lower banks in which the PQs were established (3 PQs).

Results and discussion

In Table 1 Qsm, Nm, fluctuations in soil moisture content and salinity of the soil moisture, stability of the substrate (i.e. degree of unchangeability of the substrate in the course of time – viz. one year –), and maximum cover percentage of phanerogams of all PQs have been ranked. Aquatic PQs were separated from semi-terrestrial and terrestrial PQs. The Spearman rankcorrelation has been used for testing the hypothesis of independence between all parameters mentioned in Table 1. The results have been recorded in Table 2.

The salt-marsh habitat is situated in between land and water. The vegetation is influenced by many ecological factors, the most important of which is the influence of the tides, here expressed as the frequency of tidal flooding (Fig. 1). During intermediate and neap tides the marsh is only partly flooded and at springtides the marsh is almost entirely inundated. During extreme springtides beach plains, brackish dune-slacks and saline pastures are flooded. The flood water carries salts, nutrients, organic material, sediments and algae, remaining partly on and in the salt-marsh soil as the water retires. The influence of the saline flood water manifests itself most strongly at the lower marsh levels. Precipitation and evaporation (Fig. 2) have a marked influence on the higher levels. Between these two extremes lies a gradient situation of environmental and vegetational changes.

Soil-moisture content and salinity of the soil moisture

Table 2. Spearman rank-correlation coefficients for parameters studied in permanent quadrats in the SW Netherlands. Qsm = average similarity, Nm = average number of species, A = fluctuations in soil moisture content and salinity of the soil moisture, B = stability of the substrate, C = maximum cover percentage of phanerogams.
n.s. = not significant (P > 0.05); x = P < 0.05; xx = P < 0.01; xxx = P < 0.001.

Aquatic PQs

1968-1969	Nm	A	B	C	
Qsm	0.6871 x	0.4598 n.s.	0.1976 n.s.	0.0087 n.s.	n = 12
Nm	-	0.3077 n.s.	0.2797 n.s.	0.2797 n.s.	
A		-	0.3811 n.s.	0.1294 n.s.	
B			-	0.7483 xx	

1969-1970	Nm	A	B	C	
Qsm	0.4405 n.s.	0.3929 n.s.	0.4881 n.s.	0.5928 n.s.	n = 8
Nm	-	0.5655 n.s.	0.2738 n.s.	-0.1190 n.s.	
A		-	0.2679 n.s.	-0.2738 n.s.	
B			-	0.6948 x	

1970-1971	Nm	A	B	C	
Qsm	0.2000 n.s.	0.8000 n.s.	0.2875 n.s.	-0.4000 n.s.	n = 5
Nm	-	0.6000 n.s.	0.7250 n.s.	0.2000 n.s.	
A		-	0.2250 n.s.	-0.3000 n.s.	
B			-	0.2250 n.s.	

Terrestrial PQs

1969-1970	Nm	A	B	C	
Qsm	0.9423 xxx	0.6703 x	0.6497 xx	0.7157 xx	n = 13
Nm	-	0.6181 x	0.7418 xx	0.8242 xx	
A		-	0.6147 x	0.5673 x	
B			-	0.8558 xxx	

1970-1971	Nm	A	B	C	
Qsm	0.9196 xx	0.6250 n.s.	0.7857 x	0.3036 n.s.	n = 7
Nm	-	0.5446 n.s.	0.9018 xx	0.2232 n.s.	
A		-	0.5625 n.s.	0.6786 n.s.	
B			-	0.5446 n.s.	

can be used as parameters in which 'aquatic' and 'terrestrial' influences are clearly expressed.

In the Springersgors salt marsh the moisture content of the clayish soil was high (60–70%) up to just below MHW, showing only small fluctuations; a sandy profile, however, showed larger fluctuations (40–70%). Between MHW and MHWS the soil moisture content was still high (50–70%) but its fluctuation showed a weak correlation with the evaporation surpluses mentioned. Around MHWS the soil-moisture content fluctuated considerably (20–60%), with lowest values during the dry seasons. Above MHWS two different habitat-types could be roughly distinguished in the salt marsh: (a) in saline pastures overgrown with a dense cover of halophytes (dominance of *Puccinellia maritima*, *Plantago maritima*, *Glaux maritima*) the soil-moisture content fluctuated considerably (20–60%) but was never lower than 20%: the vegetation of higher plants

112

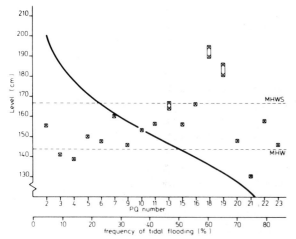

Fig. 1. Frequency of tidal flooding (sigmoid curve) in relation to the height of the permanent quadrats (small squares and rectangles) in the Springersgors. Data on tidal movements supported by Rijkswaterstaat (State Department of Roads and Waterways). Levelling of the PQs was done by the Delta Institute, Yerseke.

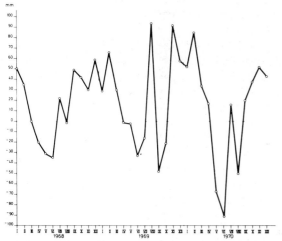

Fig. 2. Precipitation and evaporation surpluses measured at Vlissingen. The vertical axis indicates precipitation minus evaporation. Data derived from Anonymous 1968–1970.

acted as a check on evaporation; (b) in the sandy habitat at the foot of the dunes the soil was permanently dry with a moisture content mostly below 20 %, sometimes even down to 1–2 % during dry periods.

At the borders of the inland brackish water-bodies the soil-moisture content of the terrestrial PQs, which were never inundated, followed a regime, almost completely reflecting precipitation and evaporation surpluses (Fig. 2). However, it never dropped below 20 % (fluctuation

20–60 %). The sandy clay profile and the sods of phanerogams prevented it from extreme desiccation. The intermittently inundated inland PQs formed a semi-terrestrial habitat with a soil moisture content never below 35 % (fluctuation 35–70 %).

Just below MHW, in the salt-marsh habitat, the salinity of the soil moisture showed only minor fluctuations ($10–20 \, ^o/_{oo} \, Cl^-$), following the chlorinity of the open water, with very small increases in periods with marked evaporation surpluses. Between MHW and MHWS the salinity of the soil moisture fluctuated considerably, reaching values up to $100 \, ^o/_{oo} \, Cl^-$, and these fluctuations could be correlated with evaporation and precipitation surpluses. PQs situated on clay soil showed smaller fluctuations than PQs on sandy clay. Around MHWS large fluctuations were measured in the salinity of the soil moisture; far above this level salinity values fluctuated enormously in sandy soil (0 up to more than $200 \, ^o/_{oo} \, Cl^-$) and closely reflected the precipitation and evaporation surpluses (Fig. 2). In semi-terrestrial and terrestrial saline pastures large fluctuations occurred in the salinity of the soil moisture ($1–170 \, ^o/_{oo} \, Cl^-$). In terrestrial PQs along brackish inland waters, the salinity of the soil moisture fluctuated extremely and showed very low values in winter (down to $0.2 \, ^o/_{oo} \, Cl^-$) and very high values in summer (up to $200–300 \, ^o/_{oo} \, Cl^-$).

Benthic algae in salt marshes and saline pastures grow loose-lying on, or embedded in the soil (excluding the epiphytes). Attachment organs are often lacking or reduced. The stability of the superficial soil layers depends on (1) tidal floodings depositing or removing layers of sediment and algal material, (2) wave action disturbing the upper layer of sediment and attached algae, (3) eroding activities of wind and rain, (4) trampling caused by cattle and sheep, and (5) the presence of an algal mat or a stand of phanerogams.

In the salt marsh below and around MHW transport of algal material from and into the marsh occurs frequently. Waves and currents have a free entrance and silt is deposited regularly. The substrate is instable. Up to around and above MHWS the tidal influences diminish gradually. Transport of algal material becomes inconspicuous, except during storms. Higher plants stabilize the soil. The substrate is relatively stable. On bare sandy spots wind and rain frequently disturb the superficial sediment deposits (cf. Nienhuis 1973). In moderately grazed saline pastures the sods of grasses and other halophytes fix the soil, thus promoting its stability. The bare strip between water and pasture along inland waters is influenced by wave action;

the soil is weak and may be heavily trampled, even by very few cattle.

Environmental factors change and so does the algal vegetation in the PQs. It appeared that the average number of species in the relevés and the average similarity between the numbers of species of successive relevés is higher in semi-terrestrial and terrestrial PQs than in aquatic ones. The highest values were found in non-tidal saline pastures (Table 1).

In the aquatic environment of the Springersgors salt marsh proper, no significant correlation was found between the vegetational parameters (Nm and Qsm) and the environmental parameters (combined fluctuations in soil-moisture content and salinity of the soil moisture and stability of the substrate) (Table 2). Nm and Qsm were significantly correlated in 1968–1969 and were not in 1969–1970 and 1970–1971. A varying correlation was found between the stability of the substrate and the maximum cover percentage of the phanerogams (Table 2). A significant correlation between the two parameters mentioned indicates that higher plants fix the soil, thus promoting its stability.

With regard to semi-terrestrial and terrestrial environments other conclusions than on aquatic environments can be drawn from the rank-correlation scheme (Table 2). In both these environments Qsm and Nm showed a highly significant correlation. Fluctuations in soil-moisture content and in salinity of the soil moisture as well as the maximum cover percentage of the phanerogams did have either a significant (1969–1970) or a non-significant correlation (1970–1971) with Nm and Qsm. Either a very highly significant correlation or no significant correlation at all was calculated between the combined fluctuations in moisture content and salinity of the soil moisture and (1) maximum cover percentage of phanerogams, and (2) stability of the substrate. This means that both in unstable and in stable soils large fluctuations in salinity and moisture content occurred, and that, moreover, both unstable and stable substrates carried a dense stand of phanerogams. What cannot be derived from these data is the fact that on unstable soils only annual species flourished well (*Salicornia europaea*), whereas on stable soils perennial halophytes were dominant, e.g. *Limonium vulgare* and *Halimione portulacoides*.

Hardly any correlation existed between the combined fluctuations in soil-moisture content and salinity of the soil moisture on the one hand and the maximum cover of phanerogams on the other. Under semi-terrestrial and terrestrial conditions the cover of the phanerogams in the

PQs never rose above 80% (Nienhuis 1975), and the halophytic vegetation did not prevent the soil from desiccation.

In semi-terrestrial and terrestrial environments Nm and Qsm showed a significant or highly significant positive correlation with the stability of the substrate, which means that the number of species increases and the changeability of the algal flora decreases with increasing stability of the substrate. Of all parameters mentioned in Table 2 (excluding Qsm) the stability of the substrate has the most clear-cut positive correlation with the number of species.

Species diversity tends to be low in ecosystems subjected to strong physicochemical limiting factors (Odum 1971). The limiting factors discussed here are the soil-moisture content, the salinity of the soil moisture and the instability of the substrate. On account of the data presented here instability of the substrate must be regarded as an important limiting factor with respect to the number of algal species per unit area under semiterrestrial and terrestrial conditions. Fluctuations in soil-moisture content and in salinity of the soil moisture are most probably of minor importance.

The number of algal species per unit area has been taken as a measure of species diversity. However, it is important to recognize that overall species diversity usually is thought to be composed of at least two components, the species richness component (= number of species per unit area) and the 'equitability' component (= number of individuals or cover degree per species per unit area). Generally a positive correlation exists between the number of species per unit area and overall species diversity (Odum 1971).

The usefulness of quantitative measures of sample similarity in which both species richness and equitability play a role has been discussed by Goodall (1973). Several differing indices and numerous differing opinions about their values exist. The qualitative Sörensen index, on the contrary, has been used widely for expressing similarity on the basis of numbers of species.

Minor disadvantages adhere to the rank-correlation approach. The stability of the substrate is based on a subjective estimate and, moreover, in some 'aquatic' PQs small pieces with a stable substrate (woody stems of phanerogams) were present, which may have led to an increase in the number of species.

Nm and Qsm showed a significant or highly significant positive correlation with the stability of the substrate and hardly any correlation with fluctuations in soil-moisture content and salinity of the soil moisture. Nm and Qsm were based on qualitative data, and the question arises whether the roughly quantitative data of the relevés confirm these conclusions.

The maximum cover percentage of the phanerogams was used as a vegetational parameter in the rank-correlation (Table 1, 2). Halophytes show their maximum cover in summer and their minimum in winter. In contradistinction to this the maximum and minimum cover percentages of the benthic algae may show one or several maxima and minima a year. Therefore the relation between the fluctuation in soil-moisture content and the salinity of the soil moisture and the stability of the substrate on the one hand and the cover degree of the algal mat on the other one, will be discussed without using the rank-correlation method.

Small and moderate fluctuations in soil-moisture content and in salinity of the soil moisture (Table 1) had hardly any limiting influence on the algal cover degree. Only in PQ Sp 4 a decrease in cover of *Fucus vesiculosus* f. *volubilis* in April-May 1968 might probably be explained as a result of drought.

Larger fluctuations in soil moisture and salinity (Table 1) may influence the algal vegetation sharply. Bleaching of the Chlorophycean algal mat (dominated by *Enteromorpha prolifera*, *E. torta*, *Rhizoclonium riparium*, *Percursaria percursa*), followed by a decrease of the algal cover quite often coincided with extreme low moisture contents and extreme high salinities (Sp 2: June-July 1970; Sp 22: July-August 1969, June-July 1970; Bat 2: July-August 1969, June-July 1970; DW 1: July-August 1969, June-July 1970; Suz 1: June-July 1969, May-June 1970). Soil-moisture content and salinity of the soil moisture were often negatively correlated, and hence it was impossible to decide whether high salinity or low moisture or both factors were responsible for the decrease of the algal mat.

Quick desiccation of the superficial millimeters of the soil as observed in many PQs in June 1970 resulted in bleaching and drying up of the algal mat, notwithstanding the fact that the soil-water content remained at a relatively high level (Sp 13, Bat 1: ca. 30%). The upper layer of the green algal mat which was exposed to direct sunlight bleached presumably before it could be fed by available soil-water.

A decrease of the soil-water content to values below 20%, lasting for several months, and the concomitant increase of the salinity of the soil moisture to values of ca. 60–200 $^o/_{oo}$ Cl$^-$ were fatal for the stands of green algae. Their cover percentage diminished rapidly till only minor traces were left (Sp 19: June 1968 - November 1968, April 1969 - October 1969; Sp 22: July 1969 - October 1969). In PQ Sp 18 the soil-moisture content only rarely reached

values above 20%, and amounted to ca. 10% on an average. The salinity of the soil moisture in this PQ reached values of 130–190 $^o/_{oo}$ Cl$^-$, during periods with an evaporation surplus. A green algal mat was lacking there. Only *Cyanophyceae* were found all year round in small quantities (cover less than 5%). It is obvious that the soil-moisture content acted here as a limiting factor for the growth of salt-marsh green algae. Salinities of 130–190 $^o/_{oo}$ Cl$^-$ were also measured in several other PQs, without notable influence on the coverage of the algal mat (Sp 13, Sp 22, Bat 1, DW 1).

It can be concluded that extremely high salinities and concomitant extremely low moisture contents have a limiting influence on the cover degree of the green algal mat. Below a moisture content of 20% a mat of *Chlorophyceae* and *Vaucheria* spp. cannot develop.

In general, substrates with little stability (Table 1: Sp 4, Sp 21, Sp 23, Suz 2, Via 3, DW 2) only bore ephemeral growths of green algae (in spring and early summer) or blue-green algae (in summer), and not a perennial algal mat. Substrates in the tidal PQs with large stability (Table 1: Sp 13, SP 16, Bat 1, Bat 2) generally showed a perennial algal mat with an average cover of more than 50% (mainly *Chlorophyceae* and *Vaucheria* spp., in late summer overgrown by *Cyanophyceae*), unless another factor, viz. moisture content, was limiting (cf. Sp 19). Non-tidal substrates with large stability (Suz 1, Via 1, Via 2) showed an irregular picture with maximum cover percentages in spring and summer and minima in winter.

It can be concluded that substrate stability is weakly positively correlated with the stability of the algal mat, i.e. the degree of unchangeability of the algal mat in the course of time.

It is a well-known fact that severe frost lasting for several weeks may damage algal vegetation (cf. Den Hartog 1959). The influence of frost on the salt-marsh algae could hardly be studied during the period of investigation of the PQs. The winters of 1968-1969 and 1969-1970 were moderate and that of 1970-1971 was even mild. Only February 1969 had two weeks of severe frost (i.e. air temperatures down to − 10 °C). In a number of inundated PQs in the Springersgors the formation of ice was noticed during that period. In the higher situated PQs the soil and the algal mat were frozen.

The cover degree in the aquatic PQs (Table 1) was not influenced by frost. In several semi-terrestrial PQs (Table 1) the frost appeared to be harmful as could be seen from the sudden decrease in cover percentages (Sp 13, Sp 16, Sp 19). However, the frozen blue-green algae and green algae,

Table 3. List of algal species found in the PQs. Column A: present in the salt marsh (x);
B: present in inland waters (x); C: maximum cover percentage greater than or
equal to 5% (x).
Nomenclature of Chlorophyceae (excluding Rhizoclonium riparium sensu Nienhuis,
1975), Phaeophyceae (excluding Fucus vesiculosus f. volubilis sensu Powell,
1963),Haptophyceae and Rhodophyceae according to Parke & Dixon (1968). Vaucheria
species were derived from Simons, 1974. Cyanophyceae: coccoid blue-green algae
were identified with Drouet & Daily (1956). Most Oscillatoriaceae were identified
with Geitler(1932), except in mixed samples in which the delimitation between several Geitlerian species appeared to be impossible; in those cases Drouet (1968)
was used.

	A	B	C		A	B	C
Coccoid green alga	x	x		Anacystis aeruginosa	x	x	
Ulothrix flacca	x	x		Agmenellum quadruplicatum	x	x	
Ulothrix pseudoflacca	x	x	x	Agmenellum thermale		x	
Ulothrix subflaccida	x	x		Entophysalis deusta	x		
Hormidium flaccidum	x			Entophysalis conferta	x		
Monostroma oxyspermum	x			Anacystis spec.	x		
Blidingia minima	x			Spirulina subsalsa	x	x	
Blidingia marginata	x		x	Spirulina subtilissima	x	x	
Enteromorpha flexuosa	x			Spirulina spec.		x	
Enteromorpha torta	x	x	x	Spirulina tenerrima	x	x	
Enteromorpha prolifera	x	x	x	Schizothrix calcicola	x	x	x
Enteromorpha intestinalis	x	x	x	Schizothrix tenerrima	x		
Enteromorpha ralfsii	x			Phormidium tenue	x		
Enteromorpha ahlneriana	x			Phormidium fragile	x		
Enteromorpha clathrata	x			Phormidium valderianum	x		
Enteromorpha spec.	x			Phormidium foveolarum	x		
Percursaria percursa	x	x	x	Phormidium spec.	x	x	
Chaetomorpha capillaris	x			Microcoleus chthonoplastes	x	x	x
Chaetomorpha spec.	x			Oscillatoria bonnemaisonii	x	x	x
Cladophora liniformis	x		x	Oscillatoria nigroviridis	x		
Cladophora spec.	x	x		Oscillatoria limosa	x		
Rhizoclonium riparium	x	x	x	Oscillatoria margaritifera		x	x
				Oscillatoria jasorvensis		x	
Fucus vesiculosus f. volubilis	x		x	Oscillatoria tenuis	x		
cf. Acinetospora crinita	x		x	Oscillatoria brevis	x		
Ectocarpacean alga	x			Oscillatoria spec.	x	x	
				Lyngbya aestuarii	x	x	x
Apistonema carterae	x	x		Lyngbya semiplena	x	x	x
Haptophyceae	x	x		Lyngbya confervoides	x	x	
				Lyngbya lutea	x		
Vaucheria spec.	x	x	x	Lyngbya spp. ⌀ 1-3μm	x		
Vaucheria compacta	x			Microcoleus lyngbyaceus		x	x
Vaucheria coronata	x	x	x	Microcoleus vaginatus	x	x	
Vaucheria intermedia	x		x	Plectonema battersii	x		
Vaucheria thuretii	x	x	x	Anabaena torulosa	x	x	x
Vaucheria arcassonensis	x			Anabaena variabilis		x	
Vaucheria debaryana		x		Anabaena spec.	x	x	
Vaucheria litorea		x	x	Nodularia harveyana	x	x	
Vaucheria sescuplicaria		x		Hydrocoleum lyngbyaceum	x		
Vaucheria erythrospora		x		Calothrix scopulorum	x		
				Calothrix pulvinata	x	x	
Bostrychia scorpioides	x		x	Calothrix spec.	x		
				Nostoc spec.	x		
Coccochloris stagnina	x	x	x				
Anacystis montana	x			Total number	74	45	26
Anacystis dimidiata	x	x					

including *Rhizoclonium riparium*, proved to be alive and were able to grow under culture conditions. Obviously, the species survived this short lasting frost period but their cover diminished. How long these algae can stand low temperatures remained an unanswered question.

In Table 3 the algal species found in the PQs are listed. The number of species identified in tidal salt-marsh PQs is about twice the number found in inland water PQs, which does not mean, however, that most of the species only mentioned for tidal salt marshes exclusively occur in that habitat (cf. Nienhuis 1975). About 30% of the algal species had a maximum cover percentage greater than 5%. The rest of the species only occurred in small quantities, but most of them had a wide distribution (Nienhuis 1975).

Up to now only little attention has been paid to the temporal aspects of the ecology of algae inhabiting soft substrates. Carter (1932, 1933a, b) described the seasonal periodicity of benthic algae in Norfolk marshes (U.K.) in general terms: In the lower salt marshes *Ulothrix flacca* developed in January on the soil and on dead remains of plants and it flourished distinctly in February. In March *Ulothrix* began to decrease in vigour and *Enteromorpha* spp. made their appearance. These algae grew out to a carpet that lasted until September, when the soil became bare or supported only *Cyanophyceae*, except where *Vaucheria* was conspicuous. Fucoid and *Catenella-Bostrychia* communities were present the whole year round. *Blidingia minima* formed a perennial growth on woody stems of halophytes *Rhizoclonium riparium* was common during the whole year. Carter (l.c.) noticed that a period of drought associated with neap tides tends to hasten the disappearance of many forms.

In the higher salt marshes Carter (l.c.) could not detect a distinct seasonal periodicity; the area was overgrown with higher plants and algae were not abundant. The combination *Microcoleus chthonoplastes* and *Lyngbya aestuarii* occurred abundantly in the higher zones, especially in summer and autumn.

Carter's study was purely a floristic one. She mentioned only a few environmental parameters and, consequently, she was not able to correlate correctly variations in the algal vegetation with abiotic factors.

Chapman (1939) worked also in the salt marshes of Norfolk and confirmed the periodicity scheme of Carter (1933a, b). He went much deeper into the abiotic environmental factors of the salt-marsh habitat (Chapman 1938, 1939, 1940, 1941, 1959). He stated that a number of major, non-climatic factors are responsible for the appearance and disappearance of algal and phanerogamic communities in salt marshes, viz. (1) frequency, height and duration of tidal floodings, (2) precipitation and evaporation, (3) salinity of the soil moisture, (4) moisture content of the soil, (5) exchangeable calcium in the soil, and (6) the composition of the soil. Chapman (l.c.) published data on the annual fluctuations in some of these factors for several levels of the salt marsh. In some phanerogamic communities permanent quadrats were established and remapped once a year.

Chapman did not give a direct correlation of local algal distribution and periodicity with environmental parameters. The explanation of the wax and wane of algal populations was based on many assumptions and only little evidence.

In a general way my data of the SW Netherlands fit in very well with those of Carter and Chapman. The algal vegetation in these estuarine salt marshes closely resembles that of Norfolk, which is situated at about the same latitude.

In order to get a better insight into the dynamics of the salt-marsh ecosystem I introduced the PQ method in the study of algae on soft substrates: the temporal changes in the algal vegetation were analysed in close correlation with an analysis of environmental parameters (see Nienhuis 1975, and preliminary papers Nienhuis 1969, 1972, 1973; Nienhuis & Simons 1971).

The thus initiated PQ method was successfully applied by some other Dutch algologists (De Jonge 1976; Polderman 1974, 1975a, b; Polderman & Prud'homme van Reine 1973; Simons & Vroman 1973). Unfortunately only a few data on environmental parameters were available to these authors, and hence correlations between environmental variation and variation in species composition and cover degree of the algal mat were mainly based on suppositions. Polderman's merits are obvious in his thorough description of the structure and periodicity of algal vegetations in the Waddenzee salt marshes. De Jonge (1976) applied a more objective analysis to his data of the Waddenzee by defining his vegetation units on the basis of numerical processing procedures.

Summary

Knowledge of the spatial structure of benthic algal vegetations in salt marshes, consisting of about 100 species of blue-green, green, brown and red algae, formed the basis of a detailed study into the dynamics of the algal mat.

The temporal changes in algal vegetations and concomitant processes in their environment were studied in 27 permanent quadrats (PQs), plotted in tidal salt marshes and along brackish inland waters in the SW Netherlands. Once a month vegetation and environment of the PQs were examined over the period March 1968-February 1971. From the Spearman rank-correlation between vegetational and environmental parameters it appeared that in semi-terrestrial and terrestrial environments the average number of algal species in a relevé and the average similarity between the species composition of successive relevés obtained from one PQ have a significant positive correlation with the stability of the substrate, and hardly any correlation with fluctuations in soil-moisture content and salinity of the soil moisture and with the maximum cover percentage of the phanerogams.

Stable substrates tend to bear stable algal mats, as appeared from quantitative data. Small and moderate fluctuations in soil-moisture content and salinity of the soil moisture have no limiting influence on the algal cover degree. During periods of extremely high salinity and concomitant extremely low soil-moisture contents that coincide with evaporation surpluses, the green algal mat bleaches and decreases in cover in semi-terrestrial and terrestrial environments. Below 20% the soil-moisture content is limiting for the expansion of a mat of green algae (*Rhizoclonium riparium*, *Percursaria percursa*, *Enteromorpha torta*, *E. prolifera*) and *Vaucheria* species. The mat desiccates and bleaches but recovers after increase of the soil-moisture content above 20% within one month. Two weeks of severe frost (temperature down to $-10\,°C$) do not influence the cover degree of the algal mat in aquatic brackish PQs, but the cover degree in semi-terrestrial PQs decreases. Short lasting soil-moisture salinities of $130-190\ ^o/_{oo}\ Cl^-$ have no limiting effect on the expansion of the algal mat.

References

Anonymous. 1968–1970. Maandelijks overzicht van de weersgesteldheid. Koninklijk Nederlands Meteorologisch Instituut 65e–67e jaargang. Uitgave 94a.

Beeftink, W.G. 1977. The coastal salt marshes of western and northern Europe: an ecological and phytosociological approach. In: V.J. Chapman (ed.), Wet Coastal ecosystems. pp. 109–155. Elsevier, Amsterdam.

Carter, N. 1932. A comparative study of the alga flora of two salt marshes. I.J. Ecol. 20: 341–370.

Carter, N. 1933a. A comparative study of the alga flora of two salt marshes. II. J. Ecol. 21: 128–208.

Carter, N. 1933b. A comparative study of the alga flora of two salt marshes. III. J. Ecol. 21: 385–403.

Chapman, V.J. 1938. Studies in salt-marsh ecology, sections I to III. J. Ecol. 26: 144–179.

Chapman, V.J. 1939. Studies in salt-marsh ecology, sections IV to V.J. Ecol. 27: 160–201.

Chapman, V.J. 1940. Studies in salt-marsh ecology, sections VI and VII. J. Ecol. 28: 118–152.

Chapman, V.J. 1941. Studies in salt-marsh ecology, section VIII. J. Ecol. 29: 69–82.

Chapman, V.J. 1959. Studies in salt-marsh ecology IX. Changes in salt-marsh vegetation at Scolt Head Island. J. Ecol. 47: 619–639.

Doing Kraft, H. 1954. l'Analyse des carrés permanents. Acta Bot. Neerl. 3: 421–424.

Drouet, F. 1968. Revision of the classification of the Oscillatoriaceae. Acad. Nat. Sci. Philad. Monograph 16: 1–370.

Drouet, F. & W.A. Daily. 1956. Revision of the coccoid Myxophyceae. Butler Univ. bot. Stud. 12: 1–222.

Geitler, L. 1932. (reprinted in 1971). Cyanophyceae. In: L. Rabenhorst, Kryptogamen-Flora von Deutschland, Österreich und der Schweiz 14: 1–1196. Johnson Repr. Corp., New York, London.

Goodall, D.W. 1973. Sample similarity and species correlation. In: R.H. Whittaker (ed.), Ordination and classification of communities. Handbook of vegetation science 5: 107–156. Junk, Den Haag.

Hartog, C. den, 1959. The epilithic algal communities occurring along the coast of the Netherlands. Wentia 1: 1–241.

Jonge, V.N. de. 1976. Algal vegetations on salt-marshes along the Western Dutch Wadden Sea. Neth. J. Sea Res. 10: 262–283.

Nienhuis, P.H. 1969. Enkele opmerkingen over het geslacht Enteromorpha Link, op de schorren en slikken van Z.W.-Nederland. Gorteria 4: 178–183.

Nienhuis, P.H. 1970. The benthic algal communities of flats and salt marshes in the Grevelingen, a sea-arm in the south-western Netherlands. Neth. J. Sea Res. 5: 20–49.

Nienhuis, P.H. 1972. The use of permanent sample plots in studying the quantitative ecology of algae in salt marshes. Proceed. VIIth Int. Seaweed Symp., Sapporo, Japan 1971. pp. 251–254.

Nienhuis, P.H. 1973. Salt-marsh and beach plain as a habitat for benthic algae. Hydrobiol. Bull. 7: 15–24.

Nienhuis, P.H. 1975. Biosystematics and ecology of Rhizoclonium riparium (Roth) Harv. (Chlorophyceae: Cladophorales) in the estuarine area of the rivers Rhine, Meuse and Scheldt. Ph.D. Thesis Rijksuniversiteit Groningen. Bronder Offset B.V. Rotterdam. 240 pp.

Nienhuis, P.H. & J. Simons. 1971. Vaucheria species and some other algae on a Dutch salt marsh, with ecological notes on their periodicity. Acta Bot. Neerl. 20: 107–118.

Odum, E.P. 1971. Fundamentals of ecology. Saunders, Philadelphia. 574 pp.

Parke, M. & P.S. Dixon. 1968. Check-list of British marine algae – second revision. J. mar. biol. Ass. U.K. 48: 783–832.

Polderman, P.J.G. 1974. The algae of saline areas near Vlissingen (The Netherlands). Acta Bot. Neerl. 23: 65–79.

Polderman, P.J.G. 1975a. The algal communities of the north-eastern part of the saltmarsh 'De Mok' on Texel (The Netherlands). Acta Bot. Neerl. 24: 361–378.

Polderman, P.J.G. 1975b. Seasonal aspects of algal communities in salt-marshes. Colloques phytosociologiques IV. Les vases salées. pp. 479–487. Cramer, Vaduz.

Polderman, P.J.G. & W.F. Prud'homme van Reine. 1973. Chrysomeris ramosa (Chrysophyceae) in Denmark and in the Netherlands. Acta Bot. Neerl. 22: 81–91.

Powell, H.T. 1963. Speciation in the genus Fucus L. and related genera. In: Speciation in the sea. Publ. Syst. Ass. 5: 63–77.

Simons, J. 1974. Vaucheria birostris n. sp., and some further remarks on the genus Vaucheria in the Netherlands. Acta Bot. Neerl. 23: 399–413.

Simons, J. & M. Vroman, 1973. Vaucheria species from the Dutch brackish inland ponds 'De Putten'. Acta Bot. Neerl. 22: 177–192.

Sörensen, T. 1948. A method of establishing groups of equal amplitude in plant sociology, based on similarity of species content. Det. Kong. danske vidensk. selsk. Biol. skr. 5: 1–34.

SAMPLING TECHNIQUES AND DISTURBANCE OF ALGAL VEGETATION IN PERMANENT QUADRATS*,**

P.J.G. POLDERMAN***

Laboratory of Aquatic Ecology, Catholic University, Toernooiveld, Nijmegen, The Netherlands

Keywords:
Algae, Disturbance, Saltmarsh, Sampling.

Introduction

In studies of saltmarsh vegetation, the algal layers are generally not included. Recently, however, Nienhuis, Simons and Polderman consistently included the algal component in their studies on north-west European saltmarsh vegetation. They have studied the dynamics of the algal vegetation using permanent quadrats. Basically, the methods employed by Nienhuis, Simons and Polderman are similar. The permanent quadrat method of Nienhuis & Simons (1971), however, differs in some respects from that of Polderman (1975a, b). Nienhuis & Simons (1971) chose quadrats of 2 x 2 m² in size within which a 50 x 50 cm² area was used to determine fluctuations in cover and composition of the algal mat, the remaining surface being used for sampling in order to establish the floristic composition of the quadrat. Polderman (1975a, b) used quadrats of 50 x 50 cm² within which both the fluctuations in the vegetation components and the floristic composition were determined. An objection to the latter method is the risk of disturbance in the quadrat due to removing samples. This paper is concerned with sampling techniques which reduce the risk of disturbance.

*Nomenclature follows Polderman (1975b).
**Contribution to the Symposium of the Working Group for Succession Research on Permanent Plots, held at Yerseke, The Netherlands, October 1975.
***The author thanks Dr. Ir. W.G. Beeftink for helpful criticism and Mrs. R.A. Polderman-Hall for correcting the English text. The investigation of the algal communities of saltmarshes in the Wadden area is a project of the Netherlands Organisation for the Advancement of Pure Research.

Analysis procedure and consequences of sampling

Algae, unlike phanerogams, can seldom be identified in the field. In the saltmarsh, the algal vegetation consists of a more or less homogeneous mat. It can be of one uniform structure (Fig. 1) but frequently it is a mosaic of two, three or even eight components, each individually distinguishable in the field. The components are determined by one or a few dominants which are recognizable by family, generic or sometimes specific characters. Amongst these dominants companion species occur. These are not noticeable with the naked eye. Therefore, it is necessary to carry out an analysis procedure at two levels: in the field, the cover percentages of the vegetation components formed by the dominant algae have to be estimated within the quadrat, and representative samples from each component should be collected for analysis in the laboratory.

For a survey using non-permanent quadrats, usually samples with algae and a thin layer of the substratum, measuring 2–4 cm², are taken. Samples of this size are not particularly vulnerable during transportation and are useful for cultivation purposes. In the laboratory, microsamples, tiny quantities of algae from the field samples, taken with a pair of tweezers, are inspected with the aid of a microscope at several magnifications. Microsampling is continued until no more new species are found. Generally, three microsamples are sufficient to find 90% of the number of species present. In this way, species are identified and cover percentages are established more precisely. The analysis results of one vegetation unit are called the micro-survey and the overall analysis result is the macro-survey (see Table 1). Table 1 gives an example of the means by which the cover of

Fig. 1. A homogeneous mat of *Rhizoclonium riparium* occurring in the *Juncetum gerardii* of the saltmarsh 'De Dellewal' (Terschelling).

Table 1. Example of the calculation of cover percentages in an algal vegetation of a PQ in the saltmarsh "De Mok" (Texel) from the phanerogamic zone of the Puccinellietum maritimae, surveyed on 30.12.1975.

Component	Vaucheria	Filamentous green algae	Oscillatoriaceae	Total
Cover estimated in the field (%)	10	80	10	100
Laboratory results	Percentages of the species in the components			Total
	Vaucheria microsurvey	Filamentous green algae microsurvey	Oscillatoriaceae microsurvey	Macro-survey
Vaucheria intermedia	50	1[1]	-	5[4]
Vaucheria subsimplex	50	-	-	5
Vaucheria compacta	1[3]	-	-	1
Ectocarpaceae spec.	1	-	-	1
Ulothrix pseudoflacca	1	1	-	1
Ulothrix subflaccida	1	2[2]	-	2
Blidingia minima	2	1	1	2
Enteromorpha prolifera	1	2	1	2
Ulvaria oxysperma	1	-	-	1
Percursaria percursa	1	1	-	1
Rhizoclonium riparium	2	100	1	80
Planktonema lauterbornii	1	-	1	1
Anacystis dimidiata	1	1	1	1
Entophysalis deusta	-	-	1	1
Spirulina subsalsa	2	1	1	2
Microcoleus chthonoplastes	1	1	50	5
Microcoleus lyngbyaceus	1	1	2	2
Oscillatoria brevis	-	2	-	2
Oscillatoria nigroviridis	1	2	1	2
Schizothrix calcicola	-	1	-	1
Symploca atlantica	-	1	50	5

1) The symbol 1 is comparable to the symbol + of the Braun-Blanquet scale.
2) The symbol 2 is comparable to the symbol 1 of the Braun-Blanquet scale.
3) Vaucheria compacta fruited in a culture on 12-1-1976.
4) The total cover percentages of Vaucheria must be considered with some reserve.

algal species is calculated from a three component algal vegetation, consisting of filamentous green algae, Oscillatoriaceae and *Vaucheria* which are the main comprehensive groups of algae in the Wadden area. The accuracy of the values found is at the 5% level (Polderman 1974). For further details see Polderman (1975a). In the laboratory a number of species, especially those belonging to the genus *Vaucheria,* cannot be identified immediately. For this group cultivation is often necessary in order to obtain the specific reproductive structures.

In a three component mosaic vegetation, like the one given in Table 1, three samples, each of 2–4 cm² (macro-samples), cause a removal of 6–12 cm² of material from the quadrat. In a vegetation of six components, 12–24cm² are removed at each sampling. In permanent quadrats this procedure is to be repeated several times during the year since sampling at monthly intervals is general pratice (Nienhuis 1975, Simons 1975, Polderman 1975b). Even then, sudden ephemeral changes may be missed as was found in an occasional two-weekly sampling of a PQ on 'De Mok' on Texel (Table 2: 1–3 in Polderman 1975b). Collecting monthly samples of 2–4 cm² could sample away a six-component vegetation within a year.

Nienhuis & Simons (1971) have solved this problem by sampling outside the quadrat in which the cover of the components is estimated (the PQ proper sensu Nienhuis 1975), extracting them from vegetation units in a larger square of 4 m² which seems similar to the eye. Nienhuis (1975) tested the reliability of this method. In areas where the present author carried out investigations on permanent quadrats (Polderman 1974, 1975a, b), homogeneous vegetations of 4 m² were seldom available. Sometimes even smaller quadrat sizes than 50 x 50 cm² had to be worked with. Therefore, it was necessary to sample within the PQ proper. It also solved the problem of how to deal with minute changes restricted to the PQ proper, such as the origin of crusts of *Apistonema* and films of *Anabaena variabilis* which sometimes are local phenomena in larger areas. Nevertheless, taking the samples directly from the quadrats in the manner mentioned above may cause a serious disturbance which might affect the results of later surveys.

Alternative techniques to avoid disturbance by sampling

A first solution would be to choose larger quadrats, for example, of 1 m² (Polderman 1974) but, as explained above, this is often impossible.

A second solution is reduction of the size of the samples. One of the conditions an appropriate sample has to fulfil is that its size has to be larger than the minimal area of the vegetation type it represents.

Nienhuis (1972) found for several vegetation types a minimal area (sensu Van der Maarel 1966) of less than 2 cm². Tests by the author showed that the quantitative minimal area (sensu Meijer Drees 1954) in vegetation units of *Vaucheria*, several filamentous green algae and Oscillatoriaceae is less than the size of one microsample, which measures 20 mm² in surface area. The qualitative minimal area equalled the size of one, two or three microsamples. Judging from these values, samples of ¹/₄–¹/₂ cm² are sufficient for a representative picture of the vegetation unit under study. Using these reduced size samples in a quadrat containing six components, 1¹/₂–3 cm² is removed per sampling. In other words, the total area of disturbance for monthly sampling is 18–36 cm², that is less than 2% of a 50 x 50 cm² PQ per year. This must be considered an acceptable rate of disturbance.

The method of applying reduced size samples involves one new problem. The *Vaucheria* component may require cultivation in order to identify its constituting species. A *Vaucheria* sample of ¹/₄–¹/₂ cm² is often not viable, so that for *Vaucheria* actually larger samples (approx. 1 cm²) are necessary. In the case of *Vaucheria*-containing vegetations the disturbance of the algal mat becomes slightly greater but the total disturbance still does not exceed 2% per year.

A method even less harmful to the algal mat is the direct removal from the PQ of the microsamples (three for each component) for microscopic analysis. In a relatively dense mat disturbance is practically nil. This method has a further advantage in that three or more microsamples can be taken from places in the component at larger distances from each other, the distance not being limited by the size of the field sample. In this method, where the samples are smaller than in the reduced size sampling, the *Vaucheria* problem still remains unsolved. It also takes some effort to transport the algal samples adequately and a third disadvantage is that, in the microsampling method, the samples have to be studied more intensively as the number of microsamples available is limited. A more intensive study is also more time consuming. As regards avoiding disturbance, microsampling is most suitable. The reduced size sampling method is quicker as regards its analysis procedure, though certain vegetation units may disappear from a quadrat because they have been sampled away, especially if they cover only a small area.

If too much soil is removed with the macrosample depressions may be induced, forming a favourable habitat for moisture dependent species such as *Vaucheria*. On the other hand, pecking for food by a colony of oystercatchers has a much more devastating effect on the surface of a PQ than the removal of a small sample (Fig. 2). The hoof-prints of sheep, or footprints of man and dogs induce much deeper depressions than a superficial macro- or micro-sampling. In permanent quadrats where trampling occurs regularly the disturbance is so great that sampling can be considered as causing negligible disturbance. Possible effects of sampling are futher annihilated by silting up or when the vegetation disappears due to seasonal influences.

This demonstrates that a certain degree of disturbance caused by sampling in a permanent quadrat should be allowed for and the advisability of modifying the sampling method in order to avoid this as much as possible must be considered, but it also indicates that the disturbance caused by sampling should not be overrated.

Summary

In studying permanent quadrats established in an algal vegetation by means of sampling in the quadrat itself, the vegetation is exposed to the risk of disturbance by the sampling. By taking small samples (microsampling or reduced size sampling) this risk is reduced considerably.

The quantitative minimal area of vegetation units of

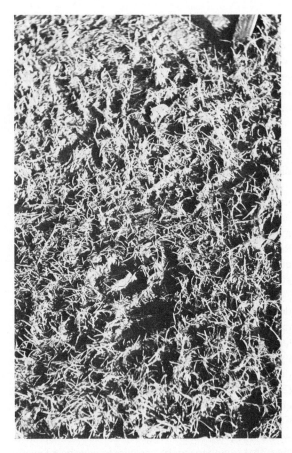

Fig. 2. Disturbance caused by oystercatchers in a permanent quadrat on 'De Mok' (Texel).

Vaucheria, filamentous green algae and Oscillatoriaceae is less than 20 mm².

In *Vaucheria* components, however, often the viability for cultivation purposes is the limiting factor for the size of a sample suitable for a complete analysis of the permanent quadrat. The minimum size of a viable *Vaucheria* sample is 1 cm².

The disturbance of the algal layer by other external factors is often more intense than that caused by sampling.

References

Maarel, E. van der. 1966. On vegetational structures, relations and systems, with special reference to the dune grassland of Voorne (The Netherlands). Diss. Utrecht. Stencilled report RIVON, Zeist. 170 pp.

Meijer Drees, E. 1954. The minimum area in tropical rain forest with special reference to some types in Bangka (Indonesia). Vegetatio 5–6: 517–523.

Nienhuis, P.H. 1975. Biosystematics and ecology of Rhizoclonium riparium (Roth) Harv. (Chlorophyceae: Cladophorales) in the estuarine area of the rivers Rhine, Meuse and Scheldt. Bronder, Rotterdam. 240 pp.

Nienhuis, P.H. & J. Simons. 1971. Vaucheria species and some other algae on a Dutch salt marsh, with ecological notes on their periodicity. Acta Bot. Neerl. 20: 107–118.

Polderman, P.J.G. 1974. The algae of saline areas near Vlissingen (The Netherlands). Acta Bot. Neerl. 23: 65–79.

Polderman, P.J.G. 1975a. Some notes on the algal vegetation of two brackish polders on Texel (The Netherlands). Hydrobiol. Bull. 9: 23–34.

Polderman, P.J.G. 1975b. The algal communities of the northeastern part of the saltmarsh 'De Mok' on Texel (The Netherlands). Acta Bot. Neerl. 24: 361–378.

Simons, J. 1975. Periodicity and distribution of brackish Vaucheria species from non-tidal coastal areas in the S.W. Netherlands. Acta Bot. Neerl. 24: 89–110.

VERBREITUNG UND DYNAMIK VON VAUCHERIA-ALGENGESELLSCHAFTEN IN DEN SÜDWESTLICHEN NIEDERLANDEN[*,**]

J. SIMONS

Biologisch Laboratorium, Vrije Universiteit, De Boelelaan 1087, Amsterdam-Buitenveldert, Niederlande

Keywords: Algal communities, Dynamics, *Vaucheria*

Einleitung

Von den 32 in den Niederlanden verbreiteten Arten der *Xanthophyceen*-Gattung *Vaucheria* sind 15 auf die Küstenregion beschränkt, wo sie besonders auf den feuchten Böden der Watten und Salzwiesen zu finden sind. Einige Arten besiedeln hier den Schlamm zu bestimmten Zeiten in solchen Massen, dass eine Beobachtung auf Dauerflächen in monatlichen Abständen lohnend erschien. Über diese Untersuchung wurde schon eingehend berichtet (Simons 1975a, b) so dass hier eine synthetische Übersicht genügt.

Untersuchungsmethode

Auf insgesamt 12 Dauerquadraten von 50 cm Kantenlänge wurde mit Hilfe eines Rasternetzes neben den Phanerogamen der Deckungsgrad der aspektbildenden Algengruppen (*Vaucheria*-Arten, fädige Grünalgen – meist *Rhizoclonium riparium* – sowie Blaualgen) mit einer fünfteiligen Skala im Gelände geschätzt. Zusammen mit der monatlichen Feldaufnahme wurden ausserhalb der Dauerfläche Bodenproben entnommen, um in Kulturversuchen nachträglich auch eine genaue Artenbestimmung vornehmen zu können.

Da für die Gezeitenvegetation der Wasser- und Salzgehalt im Boden besonders charakteristische, d.h. sowohl räumlich als auch zeitlich wechselnde Standortsfaktoren darstellen, wurde bei der monatlichen Vegetationsaufnahme in unmittelbarer Nähe des Dauerquadrats eine

weitere Bodenprobe (0–2 cm Tiefe) zur Bestimmung dieser Größen entnommen. Der Wassergehalt wurde berechnet als g H_2O/100 g Trockenboden, der Salzgehalt als g Cl^-/1 Bodenlösung ($^0/_{00}Cl^-$). (Sehe weiter Simons 1975.)

Verteilung und Dynamik der Vaucheria-Algengesellschaften im Gezeitengebiet

Für die im südwestniederländischen Gezeitengebiet typischen Pflanzengesellschaften ergibt sich eine charakteristische horizontale und vertikale Verteilung der *Vaucheria*-Arten, die in Fig. 1 schematisch wiedergegeben ist (Simons 1975a). Im *Puccinellion maritimae* und vereinzelt auch im *Armerion maritimae* der Ästuare tritt eine Gruppe von 4 Arten auf: *Vaucheria arcassonensis, V. coronata, V. intermedia* und *V. minuta*. Dabei erreicht *V. intermedia* eine höchste Dichte im Sommer und Herbst, die übrigen Arten gedeihen bevorzugt in den kälteren Jahreszeiten. Unter mesohalinen Bedingungen wird diese Gruppe allmählich durch eine andere Artenkombination ersetzt: *V. canalicularis, V. cruciata, V. erythrospora* und *V. synandra*.

Auch unterhalb der mittleren Hochwasserlinie gibt es einige auffällige *Vaucheria*-Aspekte. Durch rasches Wachstum, besonders nach vorhergehender Schlammüberlagerung, bilden sich zu bestimmten Zeiten große, einheitlich dunkelgrüne Matten. Für die marinen bis mesohalinen Zonen sind hier besonders *V. subsimplex* und *V. velutina* zu nennen, während in den Matten der meso- und oligohalinen sowie der Süßwassertideregionen *V. compacta* vorherrscht. Alle eulitoralen Arten bilden reichlich Aplanosporen aus und entfalten sich besonders im Sommer und Herbst.

Fig. 1 macht deutlich, daß die Verbreitung der *Vaucheria*-Arten eng mit dem Salzgehalt des Tidewassers korreliert ist. Auffallend ist dabei die Brackwassersubmergenz (Remane

---

[*] Nomenklatur der *Vaucheria*-Arten nach Simons (1975), der Phanerogamen nach Heukels & van Ooststroom (1970).
[**] Contribution to the Symposium of the Working Group for Succession Research on Permanent Plots, held at Yerseke, The Netherlands, October 1975.

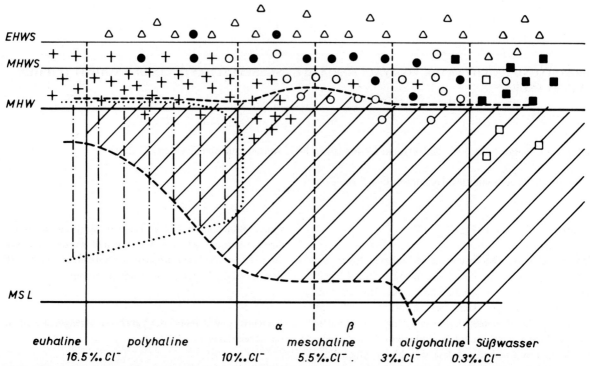

Fig. 1. Schema der horizontalen und vertikalen Verbreitung von 6 Vaucheria-Artengruppen bei unterschiedlichen Salzgehalten in den niederländischen Mündungsgebieten. Gruppe 1 (vertikale Schraffur): V. velutina. V. subsimplex. Gruppe 2 (schräge Schraffur): V. compacta. Gruppe 3 (+): V. arcassonensis, V. coronata, V. intermedia, V. minuta. Gruppe 4 (●): V. synandra, V. canalicularis, V. cruciata, V. erythrospora; oft nur V. synandra (⊖). Gruppe 5 (■): V. frigida, V. bursata, V. canalicularis, V. cruciata; oft nur V. bursata oder V. canalicularis oder beide (□). Gruppe 6 (Δ): V. terrestris, V. dillwynii. MHW: mittleres Hochwasser, MHWS: mittleres Springtide-Hochwasser, EHWS: extremes Springtide-Hochwasser, MSL: mittleres Meeresniveau.

1955) einiger Artengruppen: mit dem stromaufwärts sinkenden Salzgehalt besiedeln besonders Vertreter der Gruppen 2, 3 und 4 tiefergelegene Uferzonen, da sie erst hier die entsprechend hohen Salzkonzentrationen vorfinden, unter denen sie im höhergelegenen, küstennahen Bereich leben. Ähnliche Anpassungen fand auch Beeftink (1977) bei Phanerogamen der gleichen Region.

Verteilung und Dynamik der Vaucheria-Algengesellschaften an tidefreien Salzstellen

Die Vaucheria-Arten der Ästuare kommen auch an Salzstellen vor, die hinter den Deichen im Binnenland liegen. Allerdings ist das Vorkommen hier sehr viel unregelmäßiger als in den von der Flut erfaßten Uferzonen. Im Südwesten der Niederlande finden sich tidefreie Vaucheria-Standorte häufig an alten, quelligen Deichbruchstellen und Grabenrändern. Typisch für diese Salzstellen ist es, daß im Sommer die oberste Bodenschicht

infolge der Evapotranspiration sehr stark versalzt. In den kälteren Jahreszeiten ist der Boden so naß, daß der Salzgehalt immer sehr viel niedriger liegt. Da die Vaucheria-Arten auf feuchte Bedingungen angewiesen sind, treten sie an diesen Standorten zumeist nur im Herbst und Winter auf.

In der vertikalen Verteilung weisen die Vaucheria-Arten eine gewisse Zonierung auf: vom Wasserspiegel der Tümpel und Gräben bis zum Niveau des nicht-salinen Geländes (Deichkronen und Wiesen) folgen 4 oder 5 sich einander mehr oder weniger ausschließende Artengruppen aufeinander. Einige Beispiele, die in Simons (1975b) dokumentiert wurden seien hier kurz erwähnt. Besonders auffällig ist der Vaucheria-Aspekt im Uferbereich mit dicken, dunkelgrünen, schwammigen Polstern. V. litorea, V. sescuplicaria, V. subsimplex und V. velutina finden sich hier zusammen mit Puccinellia maritima, Salicornia europaea, Aster tripolium und Phragmites australis. Soweit V. subsimplex und V. velutina vorkommen, kann man im Sommer oft eine Massenentfaltung dieser Arten beobach-

124

ten. Ein Beispiel dafür bot im Jahr 1973 ein Graben des 'Kouderkerksche Inlaag' (Schouwen-Duiveland) mit einer raschen Zunahme von *V. velutina* ab Mai, als viele Aplanosporen keimten. Für diesen Standort war typisch, daß der Salzgehalt über das Jahr relativ hoch und konstant blieb.

Bei stärker wechselnden Salzgehalten kann *V. sescuplicaria* zur vorherrschenden Algenart werden, wie es an einem sumpfigen *Aster tripolium*-Sandufer eines Brackwassertümpels beobachtet wurde. Jeweils im Frühjahr nahm *V. sescuplicaria* stark ab, besonders auffällig im Jahr 1971. Ursache hierfür ist der Anstieg in den Salzgehalten, der im Frühjahr 1971 stärker ausfiel als 1972. Ein Vergleich mit den anderen Dauerflächen im Binnenland ergab, daß der Bodenwassergehalt an dieser Stelle niemals limitierend war

für *V. sescuplicaria*. Neben *V. sescuplicaria* traten am Teichufer noch *V. compacta, V. subsimplex* und *V. velutina* auf, aber immer in geringerer Menge als in den von der Flut erfaßten Küstenzonen. Ebenso wie im Beispiel des 'Koudekersche Inlaag' zeigte *V. velutina* die höchste Dichte im Sommer bei hohen Salzgehalten.

Auf höhergelegenen Standorten im selben Gebiet mit einer dichten, grasartigen Vegetation aus *Puccinellia maritima* und/oder *Juncus gerardii* waren folgende *Vaucheria*-Arten miteinander vergesellschaftet: *V. arcassonensis, V. coronata, V. erythrospora, V. intermedia, V. sescuplicaria* und *V. synandra*. Als Beispiel ist in Fig. 2 die Vegetationsdynamik einer Dauerfläche wiedergegeben. Auffällig war das Sommermaximum von *V. intermedia*,

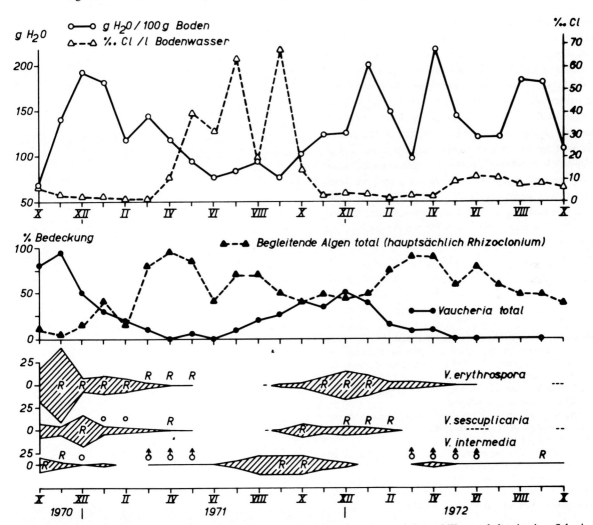

Fig. 2. Jahresdynamik der Vaucheria-Arten und der begleitenden Akgen sowie des Salz- und Wassergehaltes in einer Salzwiese. R: mit Reproduktionsorganen. O: Oosporen. ♂ : keimende Oosporen.

125

während sich die übrigen *Vaucheria*-Arten in den kälteren Jahreszeiten stärker entwickelten. Interessant ist die Wechselwirkung zwischen den *Vaucheria*-Arten, der grünen Fadenalge *Rhizoclonium riparium* und dem Salz- und Wassergehalt. *Rhizoclonium* kann hohe Salzkonzentrationen und niedrigen Wassergehalten besser widerstehen als die *Vaucheria*-Arten.

In Pflanzenbeständen, in denen *Juncus gerardii* allein vorherrscht, ist *V. synandra* mit einer kontinuierlichen geschlechtlichen Fortpflanzung zu allen Jahreszeiten die dominierende Art.

Summary

Monthly samples and observations on small permanent quadrats in the southwestern estuarine areas of the Netherlands showed that the horizontal and vertical zonation of *Vaucheria* species is clearly influenced by the salinity of the tidal water. In non-tidal saline areas salinity of the soil and particularly also the moisture-content strongly influence the occurrence of *Vaucheria* species. The dynamics of *Vaucheria* vegetation correspond to the rate of variation in soil moisture and salinity values. In tidal areas, below spring-tide level, where the tides have a tempering influence upon variation in moisture and salinity, the *Vaucheria* aspect shows more constancy than in non-tidal areas where moisture and salinity values show large seasonal variations. These examples show that the PQ-method is a good way to characterize the seasonal dynamics of algal communities.

Literatur

Beeftink, W. G. 1977. The coastal salt marshes of western and northern Europe: An ecological and phytosociological approach. In: Chapman, V. J. (ed.), Wet Coastal Ecosystems, pp. 109–155. Elsevier, Amsterdam.

Heukels, H. & S. J. van Ooststroom. 1970. Flora van Nederland, 16ᵉ druk. Noordhoff, Groningen.

Remane, A. 1955. Die Brackwasser-Submergenz und die Umkomposition der Coenosen in Belt- und Ostsee. Kieler Meeresforsch. 11: 59–73.

Simons, J. 1975a. Vaucheria species from estuarine areas in the Netherlands. Neth. J. Sea Res. 9(2): 1–23.

Simons, J. 1975b. Periodicity and distribution of brackish Vaucheria species from non-tidal coastal areas in the S.W. Netherlands. Acta Bot. Neerl. 24(2): 89–110.

ÜBER AUFGABEN UND ERGEBNISSE DER VON DER BUNDESFORSCHUNGSANSTALT FÜR NATURSCHUTZ UND LANDSCHAFTSÖKOLOGIE, BONN-BAD GODESBERG, BEARBEITETEN VEGETATIONSKUNDLICHEN DAUERFLÄCHEN*,**

A. KRAUSE

Bundesforschungsanstalt für Naturschutz und Landschaftsökologie, Heerstraße 110, D–5300 Bonn-Bad Godesberg, Bundesrepublik Deutschland

Einleitung

Die vielseitigen Aufgaben der Bundesforschungsanstalt für Naturschutz und Landschaftsökologie machen es erforderlich, daß auf zahlreichen, über das Land verteilten Flächen die Pflanzendecke zu wiederholtem Male untersucht wird.

Bei einigen Objekten stehen Standort und Bewuchs weitgehend miteinander in Einklang, bei anderen wird auf einem recht stabilen Standort eine durch menschliches Tun oder Lassen bedingte mehr oder weniger rasch verlaufende Sukzession beobachtet, und bei wieder anderen ziehen ablaufende Standortsveränderungen einen deutlich erkennbaren Umbau der Vegetation nach sich. Das Spektrum der bearbeiteten Formationen reicht von der Annuellen-Flur über Dauergrünland bis zum naturnahen Laubwald. Die Größe der untersuchten Flächen schwankt zwischen einem m² und mehreren km², und dementsprechend wechseln auch die Untersuchungsmethoden zwischen der Auszählung mit dem Gitternetz und der Wiederholungsaufnahme, im Sonderfall auch der kleinmaßstäblichen Wiederholungskartierung.

Naturwaldreservate

Die von uns bearbeiteten Naturwaldreservate (Naturwaldzellen) in Nordrhein-Westfalen sind naturnahe Waldbestände mit vorwiegend autochtoner Bestok-kung, die nahezu das gesamte Spektrum der potentiellen natürlichen Waldtypen des Landes repräsentieren. Sie wurden aus der forstlichen Nutzung herausgenommen und sollen sich selbst überlassen bleiben. Ihre angestrebte Mindestgröße beträgt 4 ha, wovon in der Regel eine 1 bis 2 ha große Teilfläche eingezäunt ist, um den Einfluß des Wildes auszuschließen (Butzke et al. 1975).

In diesen Naturwaldreservaten soll langfristigen Entwicklungsabläufen, namentlich Fragen der natürlichen Gehölzverjüngung bei bestehendem oder fehlendem Wildeinfluß nachgegangen werden. Die Vegetationsentwicklung wird mit Hilfe von Vegetationsaufnahmen der Gesamtfläche und von Teilflächen festgehalten.

Besiedlung einer neu entstandenen Flußinsel

1970 schüttete die Ahr eine Kiesinsel mit geringmächtiger Lehmdecke auf, die über das langjährige Mittelwasserniveau hinausragt und fast so hoch wie die Hartholzaue liegt. Hier wurden Dauerflächen verpflockt und ergänzend Standard-Profile ausnivelliert. Die Vegetationsentwicklung durchlief bis jetzt folgende Phasen (vgl. auch Lohmeyer 1970):

a) Im ersten Jahr wuchs eine artenreiche Therophytenflur mit über 100 Arten auf einer Fläche von 25 m² auf.

b) Im zweiten und dritten Jahr beherrschten Hochstaudenfluren die Fläche, je nach Feinerdegehalt die *Echium-Melilotus*-Gesellschaft oder ein natürliches *Tanacetum-Artemisia*-Gestrüpp.

c) Diese Hochstaudenfluren wurden abgelöst von einer Gräserflur, die auch heute noch besteht und in der *Agropyron repens* dominiert.

Gehölzaufwuchs fehlt so gut wie ganz. Schmalblatt-

*Pflanzennamen nach Oberdorfer (1970).
**Contribution to the Symposium of the Working Group for Succession Research on Permanent Plots, held at Yerseke, the Netherlands, October 1975.

weiden *(Salix purpurea, S. viminalis, S. x rubens)*, die zunächst auf dieser Fläche keimten und aufwuchsen, sind mittlerweile wieder vertrocknet. Gut entwickelt hat sich lediglich ein Exemplar von *Robinia pseudacacia*.

Entsprechende gehölzfreie Flächen gab es wohl auch früher in der Naturlandschaft. Sie dürften Siedlungsplatz für ettliche Arten gewesen sein, die heute vielfach in menschlich bedingten Graslandgesellschaften vorkommen: *Alopecurus pratensis, Arrhenatherum elatius, Poa pratensis* und *Poa trivialis*.

Natürliche generative Vermehrung von Uferweiden *(Salix* spp.)

Eine weitere Untersuchung im Anschluß an das Winterhochwasser der Ahr von 1970 galt der generativen Verjüngung von *Salix* spp. (Krause 1975). Die dreijährige Beobachtungsreihe auf abgesteckten Dauerflächen erbrachte folgende Ergebnisse:

a) Auf Sand und Kies in Wassernähe ist die *Salix*-Verjüngung unproblematisch, weil hier die Sämlinge in ihrer Entwicklung (Längenwachstum) mit den in ihrer Konkurrenzkraft geschwächten Wildkräutern Schritt halten können.

b) Auf überschlickten, stark eutrophierten Auenböden werden dagegen die *Salix*-Jungpflanzen spätestens nach wenigen Monaten von nitrophilen Hochstauden erstickt.

Natürliche Wiederbewaldung eines Kahlschlags

Auf einer Kahlschlagfläche im Buchenwaldgebiet bei Schlitz (Oberhessen) wurde über 22 Jahre lang die natürliche Wiederbewaldung registriert. Folgende Fragen standen im Vordergrund:

1. Was wächst aus Samen und aus Stöcken heran?

2. In welchem Zeitraum wird die Kahlschlagvegetation durch den zunehmenden Bestandesschatten abgelöst?

Bemerkenswert lange hielt das Vorwaldstadium mit *Rubus idaeus* und *Salix caprea* an. Dann machte sich vor allem *Prunus avium* breit. Das kann als Zeichen für eine gewisse Eutrophierung des Standortes angesehen werden, da es sich um ehemalige (spätmittelalterliche/früneuzeitliche) Ackerterrassen handelt. Bis jetzt verlief die Entwicklung jedenfalls noch nicht bis zu einem *Luzulo-Fagetum*, das die potentielle natürliche Vegetation des Gebietes darstellt.

Gartenland – Brachfläche

Auf sandigem Lehmboden eines nicht mehr bewirtschafteten Gartenlandes wurde über 8 Jahre hinweg die Vegetationsentwicklung verfolgt.

Nur im ersten Jahr war die Fläche von Therophyten – einem *Fumarietum* – besiedelt. Dann erfolgte ein völliger Umschlag, und es stellte sich ein *Ranunculus repens*-Stadium ein, das mehrere Jahre lang andauerte. Heute wird die Fläche von einer Gräser- und Staudenflur mit reichlich *Daucus carota* bedeckt.

Pflanzwald auf ehemaligem Gartenland

Auf einem Teilstück des oben genannten Gartenlandes wurde vor 8 Jahren ein nach unserer Vorstellung naturnaher Laubwald mitsamt einem aus verschiedenen Sträuchern zusammengesetzten Waldmantel gepflanzt. Auf dieser Fläche werden praktische wie wissenschaftliche Ziele verfolgt: Von Interesse für die Praxis der Gehölzverwendung im Landschaftsbau sind Auswahl des Pflanzgutes und Pflegeaufwand der Kulturfläche, von phytosoziologischem Interesse sind Fragen des Konkurrenzverhaltens im Bestand sowie der Einwanderung von Waldbodenpflanzen in die mittlerweile geschlossen herangewachsene Dickung.

Rasen-Dauerflächen an Bundesfernstraßen

Für die Begrünung von Straßenböschungen wird neben Gehölzen auch viel Rasen verwendet. Dabei wird ein pflegearmer, niedrig bleibender Magerrasen angestrebt. Die Meinungen der Fachleute über die am besten geeigneten Ansaatmischungen gehen noch weit auseinander. Deshalb wurden 1970 etwa 30 repräsentative Beobachtungsflächen an Neubaustrecken eingerichtet. Auf ihnen wird die Entwicklung verschiedener, zwischen 1963 und 1969 erfolgter Ansaaten beobachtet und bewertet (Trautmann 1972, 1975).

Auf 2 x 2 m großen Flächen wird festgestellt, welche Arten trotz Ansaat fehlen oder untervertreten sind (z.B. *Lolium perenne*), welche Arten gut und welche übervertreten sind, und schließlich, welche ohne Ansaat vorkommen. Zu letzteren gehören vor allem viele Stauden, die im Hochsommer auffällige Farbaspekte bilden: *Agrimonia eupatoria, Origanum vulgare, Centaurea scabiosa, Picris hieracioides, Hypericum perforatum,*

Linaria vulgaris und *Chrysanthemum leucanthemum.* Untersucht wird ferner, in welchem Umfang die Mutterbodenandeckung, bei der Rhizomstücke, Kriechtriebe und Samen eingebracht werden, die Rasenansaat beeinflußt.

Wie fünfjährige Beobachtungen zeigen, vollziehen sich noch immer Veränderungen im Artengefüge einiger Flächen, während sich andere bereits stabilisiert haben.

Vegetationsentwicklung auf einer frisch geschütteten Kiesfläche im Rheinischen Braunkohlenrevier

Seit 1966 wird auf einer aus unverwittertem Kies der Rhein-Hauptterrasse aufgeschütteten und eingezäunten Fläche von 0,6 ha Größe die vom Menschen unbeeinflußte Vegetationsentwicklung verfolgt. Auf 5 Ausschnitten von je 1 m² Größe finden Gitternetz-Auszählungen statt. Außerdem wird jährlich der Gesamtartenbestand der Fläche festgestellt. Hier werden Lebensformen, generative und vegetative Vermehrung, Bewurzelung, Konkurrenzverhalten und Vegetationsdynamik beobachtet.

Einige Arten hatten bereits nach drei Jahren ihre anfängliche Bedeutung verloren. So war *Senecio viscosus* im ersten Jahr optimal entwickelt, im zweiten Jahr bereits mit reduzierter Vitalität und im dritten Jahr nur noch in Zwergformen zu finden. Danach fiel diese Art quasi aus. Auch *Poa annua* verschwand nach dem dritten Jahr. Andere Arten breiteten sich zunächst schnell aus, ihr Vorrücken kam ins Stocken. So haben *Calamagrostis epigeios* und *Epilobium angustifolium* an Vitalität (Expansionsvermögen) eingebüßt. Wieder andere Arten (z.B. *Vulpia myuros* und *Hieracium bauhinii*) breiten sich gleichmäßig aus. Zahlreiche Birken (*Betula pendula*) und eine Kiefer (*Pinus sylvestris*) zeigen kontinuierlichen Zuwachs. Fast alle Gehölze keimten während der ersten drei Jahre.

Wiederholungsuntersuchungen auf Dauergrünland in Wasserentnahmegebieten

In Gebieten, in denen durch Wasserentnahme tiefgreifende Standortsveränderungen zu erwarten sind, können Untersuchungen der Pflanzendecke, namentlich des Grünlandes, gewisse Auskünfte über die Ausdehnung der beeinträchtigten Flächen geben. Hier wird die Vegetation zunächst vor den geplanten Eingriffen in den Wasserhaushalt durch ein Aufnahmenetz und eine Kartierung erfaßt. Das ganze wird dann einige Jahre nach angelaufener Wasserentnahme wiederholt.

Im Beispiel des Wassergewinnungsgeländes Lippstadt, das zuerst 1948, dann 1958 und zuletzt 1970 untersucht wurde, zeigten sich im Verlauf dieser 22 Jahre Arten- und Mengenverschiebungen und Flächenveränderungen bei den einzelnen Pflanzengesellschaften. Es war ein deutlicher Rückgang an Frische und Feuchtezeigern zu verzeichnen (u.a. *Alopecurus geniculatus, Glyceria fluitans, Lotus uliginosus, Angelica sylvestris, Filipendula ulmaria, Lychnis flos-cuculi* und *Myosotis palustris*). Gleichzeitig nahmen Arten trockener Bereiche zu oder traten neu auf (*Achillea millefolium, Bromus mollis, Festuca rubra* und *Avena pubescens*). Ein Vergleich der Vegetationskarten erlaubt es nun, die Flächen abzugrenzen, für die berechtigte Schadenersatzforderungen gestellt werden können.

Summary

The activities of the Bundesforschungsanstalt für Naturschutz und Landschaftsökologie concerning vegetation succession are presented in a concise review. They have a wide scope and are of scientific as well as of technical interest. Studies are being carried out
– in natural forest reserves to examine regeneration and competition;
– on recently raised islands of the river Ahr to investigate the pioneer plant cover and the spontaneous reproduction of riparian willows (*Salix* spp.);
– on a clearing to get knowledge on stability of pre-forest vegetation and forest regeneration out of stumps and seeds;
– on abandoned arable land to observe the changing aspects;
– in a planted forest on ancient garden ground to investigate the invasion of woodland species;
– in highway lawns to develop seed mixtures for grassy vegetation which needs little care;
– on a man-made gravel plain to observe the spontaneous colonization and vegetation dynamics;
– in meadows to analyse the effects of lowering the ground water table.

Literatur

Butzke, H. et al. 1975. Naturwaldzellen I. Schriftenreihe der Landesanstalt für Ökologie, Landschaftsentwicklung und Forstplanung Nordrhein-Westfalen 1: 1–103.

Krause, A. 1975. Über die natürliche Verjüngung von Uferweiden an der Ahr. Schr. Reihe Vegetationskunde 8: 99–104.

Lohmeyer, W. 1970. Über das Polygono-Chenopodietum in Westdeutschland unter besonderer Berücksichtigung seiner Vorkommen am Rhein und im Mündungsgebiet der Ahr. Schr. Reihe Vegetationskunde 5: 7–28.

Oberdorfer, E. 1970. Exkursionsflora für Süddeutschland und angrenzende Gebiete. Ulmer, Stuttgart.

Trautmann, W. 1972. Erste Ergebnisse von Rasenuntersuchungen an Dauerflächen der Bundesautobahnen. Natur und Landschaft 47: 70–76.

Trautmann, W. 1975. Zur Entwicklung von Rasenansaaten an Autobahnen. Natur und Landschaft 50: 45–48.

List of participants

J. P. Bakker	Laboratory for Plant Ecology, University of Groningen, Post Box 14, 9750 AA Haren, The Netherlands
Dr. W. G. Beeftink	Delta Institute for Hydrobiological Research, Vierstraat 28, 4401 EA Yerseke, The Netherlands
Dr. M. Bilio	Centro di Ricerche Ittiologiche, S.I.V.A.L.C.O., Via Mazzini 200, 44022-Comacchio, Ferrara, Italy
S. O. Borgegård	Department of Ecological Botany, University of Uppsala, Box 559, S-75122 Uppsala, Sweden
Dr. S. Bråkenhielm	Department of Ecological Botany, University of Uppsala, Box 559, S-75122 Uppsala, Sweden
Prof. Dr. H. Dierschke	Systematic-Geobotanical Institute, The University, Untere Karspüle 2, 34-Göttingen, F.R.G.
Prof. Dr. W. Eber	The University, Ammerländer Heerstrasze 67-99, 29-Oldenburg, F.R.G.
Dr. J. B. Faliński	Geobotanical Station, The University of Warsaw, 17-230 Białowicza, Woj. Białystok, Poland
M. Groenhart	Hugo de Vries Laboratory, Department of Phytosociology and Experimental Plant Ecology, Univ. of Amsterdam, Sarphatistraat 221, 1018 BX Amsterdam, The Netherlands
A. Jensen	Botanical Institute, The University of Aarhus, Nordlandsvej 68, 8240-Risskov, Denmark
Dr. W. Joenje	Laboratory for Plant Ecology, The University of Groningen, Post Box 14, 9750 AA Haren, The Netherlands
Dr. A. Krause	Bundesforschungsanstalt für Naturschutz und Landschaftsökologie, Heerstrasze 110, 5300-Bonn-Bad Godesberg 1, F.R.G.
D. van der Laan	Institute for Ecological Research, Department for Dune Research, Weevers' Duin, Duinzoom 20a, 3233 EG Oostvoorne, The Netherlands
Dr. L. de Lange	Hugo de Vries Laboratory, Department of Phytosociology and Experimental Plant Ecology, Univ. of Amsterdam, Sarphatistraat 221, 1018 BX Amsterdam, The Netherlands
A. Larsson	Department of Ecological Botany, The University of Lund, Östra Vallgatan 14, S22361-Lund, Sweden
Dr. G. Londo	Research Institute for Nature Management, Broekhuizen Castle, Leersum, The Netherlands
Dr. E. van der Maarel	Department of Geobotany, Catholic University, Toernooiveld, Nijmegen, The Netherlands
Dr. H. Muhle	Systematic-Geobotanical Institute, The University, Untere Karspüle 2, 34-Göttingen, F.R.G.
Dr. P. H. Nienhuis	Delta Institute for Hydrobiological Research, Vierstraat 28, 4401 EA Yerseke, The Netherlands
Mrs. K. van Noordwijk-Puijk	Fazantenkamp 92, 3607 CD Maarssenbroek, The Netherlands
P. J. G. Polderman	Polarisstraat 12, 3235 TH Rockanje, The Netherlands
Dr. F. Romane	Centre d'Etudes Phytosociologiques et Ecologiques Louis Emberger, Route de Mende, Post Box 5051, 34022 Montpellier, France
E. Rosén	Department of Ecological Botany, University of Uppsala, Box 559, S-75122 Uppsala, Sweden
H. L. F. Saeijs	Rijkswaterstaat Delta Department, Environmental Division, Grenadierweg 31,

Dr. H. Schmeisky

Dr. W. Schmidt

Prof. Dr. G. Schwerdtfeger and
Mrs. Schwerdtfeger

Dr. J. Simons

Dr. E. Sjögren

Miss A. W. Stienstra

Dr. D. C. P. Thalen

IJ. de Vries

Dr. E. Waldemarson Jensén

Dr. I. S. Zonneveld

Post Box 439, 4330 AK Middelburg, The Netherlands
Gesamthochschule Kassel, OE-Landwirtschaft, Nordbahnhofstrasze 1a, 3430-Witzenhausen, F.R.G.
Systematic-Geobotanical Institute, The University, Untere Karspüle 2, 34-Göttingen, F.R.G.

Am Tannenmoor 34, 3113 Suderburg 1, F.R.G.
Biological Laboratory, Free University, De Boelelaan 1087, 1081 HV Amsterdam-Buitenveldert, The Netherlands
Department of Ecological Botany, The University of Uppsala, Box 559, S-75122 Uppsala, Sweden
Delta Institute for Hydrobiological Research, Vierstraat 28, 4401 EA Yerseke, The Netherlands
International Institute for Aerial Survey and Earth Sciences (I.T.C.), 144 Boulevard 1945, Enschede, The Netherlands
Laboratory for Plant Ecology, The University of Groningen, Post Box 14, 9750 AA Haren, The Netherlands
Department of Ecological Botany, The University of Uppsala, Box 559, S-75122 Uppsala, Sweden
International Institute for Aerial Survey and Earth Sciences (I.T.C.), 144 Boulevard 1945, Enschede, The Netherlands

Contributions published in other periodicals

Sven Bråkenhielm, Permanent-quadrat studies of abandoned farm land after afforestation with Norway spruce in Southern Sweden. Data and results included in: Sven Bråkenhielm, Vegetation dynamics of afforested farmland in a district of South-eastern Sweden. Acta Phytogeogr. Suecica **63**, 1977: 1–106.

Helge Schmeisky, Sukzessionen auf Salzrasen des Graswarders vor Heiligenhafen/Ostsee. Mitteil. aus dem Ergänzungs-studium Ökologische Umweltsicherung **2**, 1977: 103–114.

Unpublished contributions

Martin Bilio, Some problems in the quantitative evaluation of fouling on experimental panels.

V. Karpáti, Einflusz intensiven Ackerbaus (Mineraldünger) auf Wasser- Ökosysteme im Naturschutzgebiet "Kisbalaton".

H. Muhle, Zur Planung von Kleinquadrat-Dauerflächen von hauptsächlich epihytischen Kryptogamen-Gesellschaften.

G. Schwerdtfeger, Dauerquadrate in einer aufgelassenen Sandgrube des Altpleistocäns der Lüneburger Heide.

Reference to original publications in VEGETATIO

W. G. Beeftink	1979. Vegetatio 40: 101–105
G. Londo	1978. Vegetatio 38: 185–190
D. C. P. Thalen	1979. Vegetatio 39: 185–190
J. B. Faliński	1978. Vegetatio 38: 175–183
P. Poissonet et al.	1978. Vegetatio 38: 135–142
I. Isépy	1978. Vegetatio 37: 187–189
W. Schmidt	1978. Vegetatio 36: 105–113
E. van der Maarel	1978. Vegetatio 38: 21– 28
D. van der Laan	1979. Vegetatio 39: 43– 51
K. van Noordwijk-Puijk et al.	1979. Vegetatio 39: 1– 13
W. Joenje	1978. Vegetatio 38: 95–102
J. P. Bakker	1978. Vegetatio 38: 77– 87
P. H. Nienhuis	1978. Vegetatio 38: 103–112
P. J. G. Polderman	1978. Vegetatio 36: 187–190
J. Simons	1978. Vegetatio 38: 119–122
A. Krause	1978. Vegetatio 36: 119–122